D1223588

Springer Tracts in Modern Physics
Volume 161

Springer
Berlin
Heidelberg
New York
Barcelona
Hong Kong
London
Milan
Paris
Singapore
Tokyo

Springer Tracts in Modern Physics

Springer Tracts in Modern Physics provides comprehensive and critical reviews of topics of current interest in physics. The following fields are emphasized: elementary particle physics, solid-state physics, complex systems, and fundamental astrophysics.

Suitable reviews of other fields can also be accepted. The editors encourage prospective authors to correspond with them in advance of submitting an article. For reviews of topics belonging to the above mentioned fields, they should address the responsible editor, otherwise the managing editor. See also http://www.springer.de/phys/books/stmp.html

Managing Editor

Gerhard Höhler

Institut für Theoretische Teilchenphysik
Universität Karlsruhe
Postfach 69 80
D-76128 Karlsruhe, Germany
Phone: +49 (7 21) 6 08 33 75
Fax: +49 (7 21) 37 07 26
Email: gerhard.hoehler@physik.uni-karlsruhe.de
http://www-ttp.physik.uni-karlsruhe.de/

Elementary Particle Physics, Editors

Johann H. Kühn

Institut für Theoretische Teilchenphysik
Universität Karlsruhe
Postfach 69 80
D-76128 Karlsruhe, Germany
Phone: +49 (7 21) 6 08 33 72
Fax: +49 (7 21) 37 07 26
Email: johann.kuehn@physik.uni-karlsruhe.de
http://www-ttp.physik.uni-karlsruhe.de/~jk

Thomas Müller

Institut für Experimentelle Kernphysik
Fakultät für Physik
Universität Karlsruhe
Postfach 69 80
D-76128 Karlsruhe, Germany
Phone: +49 (7 21) 6 08 35 24
Fax: +49 (7 21) 6 07 26 21
Email: thomas.muller@physik.uni-karlsruhe.de
http://www-ekp.physik.uni-karlsruhe.de

Roberto Peccei

Department of Physics
University of California, Los Angeles
405 Hilgard Avenue
Los Angeles, CA 90024-1547, USA
Phone: +1 310 825 1042
Fax: +1 310 825 9368
Email: peccei@physics.ucla.edu
http://www.physics.ucla.edu/faculty/ladder/
peccei.html

Solid-State Physics, Editor

Peter Wölfle

Institut für Theorie der Kondensierten Materie
Universität Karlsruhe
Postfach 69 80
D-76128 Karlsruhe, Germany
Phone: +49 (7 21) 6 08 35 90
Fax: +49 (7 21) 69 81 50
Email: woelfle@tkm.physik.uni-karlsruhe.de
http://www-tkm.physik.uni-karlsruhe.de

Complex Systems, Editor

Frank Steiner

Abteilung Theoretische Physik
Universität Ulm
Albert-Einstein-Allee 11
D-89069 Ulm, Germany
Phone: +49 (7 31) 5 02 29 10
Fax: +49 (7 31) 5 02 29 24
Email: steiner@physik.uni-ulm.de
http://www.physik.uni-ulm.de/theo/theophys.html

Fundamental Astrophysics, Editor

Joachim Trümper

Max-Planck-Institut für Extraterrestrische Physik
Postfach 16 03
D-85740 Garching, Germany
Phone: +49 (89) 32 99 35 59
Fax: +49 (89) 32 99 35 69
Email: jtrumper@mpe-garching.mpg.de
http://www.mpe-garching.mpg.de/index.html

Klaus Richter

Semiclassical Theory of Mesoscopic Quantum Systems

With 50 Figures

 Springer

Priv.-Doz. Dr. Klaus Richter

Max-Planck-Institut für
Physik komplexer Systeme
Nöthnitzer Strasse 38
01187 Dresden
Germany
E-mail: richter@mpipks-dresden.mpg.de

Library of Congress Cataloging-in-Publication Data applied for.

Die Deutsche Bibliothek – CIP Einheitsaufnahme

Richter, Klaus: Semiclassical theory of mesoscopic quantum systems / Klaus Richter. – Berlin; Heidelberg; New York; Barcelona; Hong Kong; London; Milan; Paris; Singapore; Tokyo: Springer, 2000 (Springer tracts in modern physics; Vol. 161) ISBN 3-540-66566-8

Physics and Astronomy Classification Scheme (PACS): 03.65.Sq, 05.45.Mt, 73.23.-b, 73.20.Dx, 73.23.Ad

ISSN 0081-3869
ISBN 3-540-66566-8 Springer-Verlag Berlin Heidelberg New York

Typesetting: Camera-ready copy by the author using a Springer LATEX macro package
Cover design: *design & production* GmbH, Heidelberg

Printed on acid-free paper   SPIN: 10713184   56/3144/tr   5 4 3 2 1 0

Preface

Mesoscopic physics has emerged as a new, interdisciplinary field combining concepts of atomic, molecular, cluster, and condensed-matter physics. On the one hand mesoscopic systems represent an important class of electronic devices in the rapidly growing fields of micro- and nano-physics. Quantum interference effects in these small, low-dimensional electronic systems have led to various novel physical phenomena. On the other hand mesoscopic physics has posed conceptually new questions to theory. These systems are usually too complex to treat starting from microscopic models. Moreover, they often show features of coherent quantum mechanics combined with statistical properties and classical chaos. Hence mesoscopics has developed into a prominent field of application of quantum chaos.

This book combines both mesoscopics and quantum chaos. This connection between mesoscopic quantum phenomena and classical dynamics is naturally achieved in the framework of advanced semiclassical methods. The purpose of this book is on the one hand to present basic concepts of modern semiclassical theory. On the other hand emphasis is put on the further development and adaption of these concepts to current problems in mesoscopic physics. In particular, the presentation is guided by the aim of demonstrating that semiclassical theory not only includes very suitable and inherent concepts for dealing with general problems of quantum chaos but also provides powerful tools to quantitatively compute mesoscopic quantities.

The book begins with an introductory chapter on modern semiclassical concepts in the context of mesoscopic physics. The formalism developed there serves as a platform for current research topics in mesoscopics. On the basis of the semiclassical representation of Green functions, semiclassical trace formulas in terms of classical phase-carrying (periodic) orbits are presented for mesoscopic spectral, thermodynamic, and transport quantities. In Chap. 3 semiclassical concepts for ballistic quantum transport through both phase-coherent microstructures and macroscopic patterned systems are reviewed. As examples, conductance in semiconductor billiards and magnetotransport in antidot lattices are treated and are compared with related experiments.

The semiclassical approach naturally provides a decomposition of mesoscopic quantities into a dominant, smooth classical part and additional quantum oscillations reflecting interference effects due to the confinement. Such quantum size effects are inherent in the orbital magnetism of mesoscopic systems, since there exists no classical counterpart. The geometrical effects on orbital magnetism are reviewed in detail on the level of both individual samples and ensembles. Furthermore, weak-disorder effects in quantum dots are discussed, and a semiclassical way to treat the crossover from ballistic to diffusive dynamics is outlined. Finally a semiclassical approach to electron–electron interaction effects on orbital magnetism is presented.

This book grew out of my *Habilitation* thesis at the University of Augsburg. Thus the selection of topics and the particular emphasis put on semiclassical concepts certainly reflect my working experience. I did not attempt to achieve a complete and fully balanced account of semiclassical approaches in mesoscopic physics but tried at least to mention further relevant work whenever appropriate. This also implies that several topics of the more general field of quantum chaos, with its overwhelming literature of the last decade, are not or are only briefly discussed.

In carrying out the research in this area I have benefited from a long and continuous collaboration with Rodolfo Jalabert and Denis Ullmo, which constitutes the basis of large parts of the present work. I further acknowledge fruitful collaboration with Harold Baranger, Ed McCann, Bernhard Mehlig, and Felix von Oppen on the theoretical side and Dieter Weiss on the experimental side during different stages of the work. Furthermore, I am grateful to many friends and colleagues for valuable and helpful conversations. I would like to thank Ulrike Goudschaal, Ed McCann, and Jens Nöckel for carefully reading major parts of the manuscript.

I am particularly indebted to Oriol Bohigas, Peter Fulde, Peter Hänggi, and Gert-Ludwig Ingold for their continuous support, numerous helpful discussions, and interest in the work. Moreover, I would like to thank all members of the Division de Physique Théorique at the IPN in Orsay, of the Theoretische Physik I+II at the University of Augsburg, and of the Max-Planck-Institut für Physik komplexer Systeme in Dresden for creating a friendly and stimulating atmosphere where research has been fun.

Finally I would like to thank Ms. Ute Heuser, Dr. Hans J. Kölsch, and Mrs. Jacqueline Lenz at Springer-Verlag for their help, patience, and cooperation. I acknowledge partial financial support from the Alexander von Humboldt Foundation and the DAAD through the PROCOPE programme.

Dresden, August 1999   *Klaus Richter*

Contents

1. Introduction

During the last two decades mesoscopic physics has evolved into a rapidly progressing and exciting interdisciplinary field of physics. Mesoscopic electronic systems build a bridge between microscopic objects such as atoms on the one side and macroscopic, traditional condensed-matter systems on the other side.[1] These structures, which are less than or of the order of a micron in size, represent a challenge to experimentalists, since they demand elaborate fabrication processes and involve delicate measurements. The motivation for theoreticians is not any smaller, because, on the one hand, the approaches employed for systems on macroscopic scales no longer apply or at least need refinement. On the other hand, mesoscopic structures are often too large and complex to treat them starting from a microscopic model. Moreover, as originally motivated by experiment, not only individual systems but also the response of a whole ensemble of similar systems is of interest. This has directed theoretical activities towards considerations of average quantities and their statistical properties.

Mesoscopic devices frequently exhibit both *classical*, though peculiar, remnants of bulk features, and *quantum* signatures. At low temperature, the coherence of the electron wavefunctions is retained over micron scales and hence may extend over the whole system. Thus mesoscopic behavior calls for new theoretical methods which combine statistical concepts and assumptions on the one hand with tools to treat coherent quantum mechanics on the other hand.

Examples of systems whose behavior can be classified generally as mesoscopic are found in diverse fields of physics: in nuclear scattering processes, strongly perturbed Rydberg atoms, polyatomic molecules, "quantum corrals", acoustic waves, microwaves and optical radiation in cavities, clusters, and electrons in small metallic particles or semiconductors of reduced dimensionality. The latter, mesoscopic electronic devices, will be the focus of application of the theoretical concepts presented in this book.

The emergence of quantum interference in these structures has given rise to a variety of novel, often surprising effects: universal conductance fluctua-

---

[1] For an overview showing the shrinkage of electronic components from the transistor to one-atom point contacts see [1].

tions in disordered samples, quantized conductance[2] and force oscillations [4] in microjunctions, persistent currents in rings,[3] weak localization,[4] and other Aharonov–Bohm-like effects, to name a few. These quantum phenomena constitute the heart of mesoscopic physics.[5]

Initially, disordered metals were the focus of interest in mesoscopics. The advent of high-mobility semiconductor heterostructures, the basis of the physics of two-dimensional electron gases, and advances in lithographic techniques have allowed the confinement of electrons in nanostructures of controllable geometry. These rather clean systems, where impurity scattering is strongly reduced, have been termed *ballistic* since scattering comes from specular reflection on the boundary.

The wide range of experimentally accessible systems – metal and semiconductor, disordered and ballistic, normal and superconducting – have made mesoscopic physics an interface between apparently different theoretical approaches.

There are, on the one hand, methods which have been especially designed to deal with random potentials in disordered metals. Traditionally, diagrammatic perturbation theory in a random potential has been a very useful tool [2]. During the last decade, powerful nonperturbative methods, in particular the supersymmetry method [8, 21], have attracted considerable interest and have been applied to a large number of different problems in mesoscopics.

On the other hand, approaches dealing with *quantum chaos* have been directed towards mesoscopic physics, since these methods appear promising for combining statistical concepts with quantum coherence. Moreover, phase-coherent ballistic nanostructures can be regarded as ideal laboratories for investigating chaos in quantum systems. Quantum chaos, as a novel discipline, devotes itself to the relation between classical and quantum mechanics; in particular, the question of how classical chaotic behavior is reflected on the level of the corresponding quantum system. Billiards have traditionally served as prominent model systems in quantum chaos: they combine conceptual simplicity – the model of a free particle in a box – with complexity with regard to the character of the classical dynamics and to features of the spectra and wavefunctions. Hence, the possibility of realizing such quantum billiards in microconductors has been fascinating and has opened a whole branch of research.

---

[2] For reviews on transport in disordered systems see e.g. [2]; for a review with the main focus on ballistic systems see [3].

[3] For a historical account of persistent currents see e.g. [5].

[4] For reviews of weak localization in disordered systems see e.g. [6]; for a semiclassical approach to weak localization see [7].

[5] For a number of books, general reviews, and recent special issues on mesoscopic physics see [3, 8–20].

Originally, two main approaches to quantum chaos could be distinguished: random-matrix theory and semiclassical techniques. Random-matrix theory[6] has been developed and proved very powerful for complex systems where the a priori knowledge of the Hamiltonian is rather limited.

Semiclassical techniques [26–30] probably allow one to combine classical and quantum mechanics in the most direct way. Modern semiclassical theory[7] is based on the trace formulas introduced by Gutzwiller for chaotic systems and by Berry and Tabor, as well as Balian and Bloch, for the integrable case. Semiclassical trace formulas are sums over Fourier-like components associated with classical paths and establish a connection between quantum objects such as the spectral density and pure classical terms such as the action along the orbits and stability amplitudes. Since the actions enter as phases, interference effects are introduced.

More than ten years ago, the semiclassical branch of quantum chaos began to receive considerable attention, first in atomic and molecular physics, when semiclassical Fourier techniques allowed one to unveil signatures of classical periodic orbits in the photoabsorption of Rydberg atoms [32]. In addition to the semiclassical *analysis* of experimental or quantum spectra, which has evolved into a frequently employed tool for understanding complex spectra, it has become a challenge to *synthesize* quantum spectra and compute individual energy levels on the basis of trace formulas and pure classical entities. This task implies, for chaotic systems, that one has to overcome the convergence problems of trace formulas, and it has directed interest to the question of proper resummation techniques for trace formulas of model systems [33,34] and highly excited atomic systems [35,36]. Though considerable progress has been made in dealing with trace formulas for chaotic systems, a systematic and precise semiclassical computation of energy levels remains an open problem. Related problems appear on the level of energy-level correlators when considering energy scales of the order of the mean level spacing Δ.

With the development of ballistic microstructures mesoscopic physics has emerged as a novel field of application of semiclassical methods. Ballistic microstructures are particularly suitable for mesoscopic systems for the following reasons:

(i) The Fermi wavelength λ_F, which is, for example, of the order of 40 nm in GaAs heterostructures, is usually the shortest length scale. It is, in particular, much shorter than the typical system size a, if we exclude the extreme limit of few-electron quantum dots, so-called artificial atoms [37]. In other words, the action functionals of the relevant classical paths are considerably larger than $\hbar$. Both criteria justify the application of

---

[6] For books on random-matrix theory, see e.g. [22]; for comprehensive reviews of random-matrix theory see [23, 24] and, with respect to mesoscopic quantum transport, [25].

[7] See [31] for a broad overview of this subject with about 400 annotated references, as well as the reprint book from the same reference.

semiclassical approximations and, at the same time, render numerical quantum calculations difficult, since one has to deal with highly excited, complicated wavefunctions in a single-particle picture.

(ii) Mesoscopic systems are influenced by various effects such as temperature, (weak) disorder, and electron–electron and electron–phonon interaction which introduce further characteristic length scales: the thermal length $L_T = \hbar v_F \beta/\pi$ (v_F is the Fermi velocity and $\beta = 1/k_B T$), the elastic mean free path l with respect to impurity scattering, and the phase-coherence length ℓ_ϕ which accounts for inelastic processes.

In terms of energy scales, the microscopic quantum limit is reached for $k_B T < \Delta$, where the temperature is low enough to enable the resolution of individual levels (assuming quantum coherence, $\ell_\phi \gg a$). The opposite, macroscopic limit is reached if $L_T < a$ or $k_B T > E_c = \hbar v_F/a$. For ballistic systems, E_c is the energy conjugate to the time of flight through the system. It represents the largest energy scale in the spectral density arising from the finite system size. Hence, we can speak of the mesoscopic regime if $\Delta < k_B T < E_c$ [38]. Therefore, for the study of thermodynamic spectral quantities and certain aspects of transport properties it is often not necessary to compute the spectral properties on scales below Δ. This favors semiclassical methods, since trace formulas introduce a hierarchy of energy scales according to spectral modulations related to classical paths of different length. The shortest (periodic) paths give rise to structure in the density of states on scales of E_c; the maximum path length included governs the spectral resolution. Indeed, for some of the applications to be discussed in this book only a few fundamental orbits are sufficient to describe the essential physics.

(iii) As already mentioned, ballistic mesoscopic systems are ideal tools to study the connection between classical dynamics and wave interference phenomena. In particular, it has turned out that the quantum properties of classically chaotic structures are often quite different from those of regular, nonchaotic systems. In most cases, these differences have been discussed for the density of states or, as in atomic physics, for photoabsorption. The spectral density of mesoscopic devices is usually not directly accessible, and hence other quantum quantities such as orbital magnetism and quantum transport through open systems move into the focus. One prominent area is phase-coherent transport through ballistic cavities, where the quantum conductance can be related to classical (chaotic) scattering of the electrons.[8]

The generic character of chaotic systems allows one to characterize quantum corrections to averaged quantities by a single scale, whereas integrable

---

[8] For reviews of ballistic quantum transport, including semiclassical aspects, see [40,41,43]. Classical and quantum mechanical scattering, as well as a semiclassical approach to the S-matrix, were reviewed by Smilansky in [12, 28], and are the topics of [39].

dynamics is usually reflected in a less uniform, system-specific behavior. Furthermore, differences between regular and chaotic dynamics manifest themselves on the quantum level in a different dependence on $\hbar$, which translates into a parametric difference in the magnitudes of the corresponding quantum features. We shall consider this point particularly in the context of orbital magnetism [45]. Owing to the absence of a classical magnetic moment, differences in the quantum corrections are not masked by an additional classical contribution (usually dominant in other cases).

Mesoscopic physics has intensified the interrelationships between the three major theoretical frameworks in the context of quantum chaos and has brought them closer together: supersymmetry techniques [8], random-matrix theory, and semiclassics. For disordered systems, the equivalence between random-matrix theory and the zero-dimensional σ model, deduced from supersymmetry, has already been known for some time. More recently, supersymmetry models have also been extended to ballistic chaotic systems [46–48]. By associating the diffusion operator of disordered systems with the Perron–Frobenius operator of general chaotic systems, level-density correlators of the latter could be determined. It was argued [48], moreover, that the Bohigas–Giannoni–Schmit conjecture, stating that in the classical limit the statistical spectral properties of chaotic systems coincide with random-matrix theory, has been proved using a new semiclassical field theory. This issue is still under intense discussion and is not yet settled but shows the apparent convergence of the different theoretical approaches.

1.1 A Few Examples

Before we enter into semiclassical theory we shall illustrate different facets of electronic mesoscopic quantum phenomena with the help of a few examples. The first two concern charge transport in high-mobility semiconductor heterostructures, namely experiments on "antidot crystals" and quantum dot billiards, where quantum effects in the measured conductance exhibit signatures of chaotic classical orbits. The third example represents mesoscopic enhancement of orbital magnetism in ballistic quantum dots. It is followed by a semiclassical analysis of the spectral properties of semiconductor–superconductor structures, elucidating the role of classical dynamics in the formation of a proximity gap in the density of states of the normal-conducting quantum dot. We close the list of examples with a quantum–experimental comparison of surface electron waves confined in Cu-"quantum corrals".

1.1.1 Antidot Superlattices

The role of classical periodic orbits in the context of mesoscopic conductivity has become evident in experiments on so-called antidot superlattices. These structures consist of a periodic array of nanometer-sized holes etched into semiconductor sandwich structures. This procedure, shown in Figs. 1.1a,c, results in a periodic potential landscape for the two-dimensional electron gas (2DEG) at the interface of the heterojunction. The effective potential looks similar to an egg carton (Fig. 1.1b). The electrons move at a constant Fermi energy in between the periodically arranged potential posts. If the antidots are steep, a unit cell of the antidot crystal may be regarded as an experimental realization of the Sinai billiard, one of the most prominent systems used for the theoretical investigation of classical and quantum chaos.

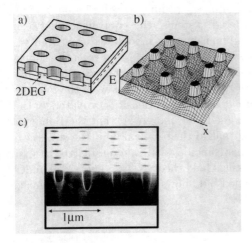

Fig. 1.1. Sketch of an array of periodically arranged holes etched into a heterojunction (**a**) and of the corresponding electrostatic antidot potential landscape (**b**) probed by the conduction electrons in the 2DEG. (**c**) The electron micrograph displays the periodically arranged holes on top of the heterojunction as well as their profile at a cleaved edge of the device. (From [52], by permission)

Such artificial lattices open up the possibility to investigate electrical transport in an interesting regime, not accessible previously: the elastic mean free path (MFP) l and the transport MFP[9] l_T are both considerably larger ($\sim$2–20 µm) than the lattice constant a. On the other hand, the Fermi wavelength $\lambda_F \sim 40$ nm is smaller than a. Hence, an antidot array can be considered as an artificial two-dimensional crystal with semiclassical electron dynamics. Since λ_F is by far the shortest length scale in the system, semiclassical transport approaches are justified.

The combined potential of the superlattice and an external magnetic field B gives rise to a variety of peculiar phenomena which will be discussed in

---

[9] This denotes the distance over which the electron momentum is randomized (see Chap. 5 and Appendix A.3).

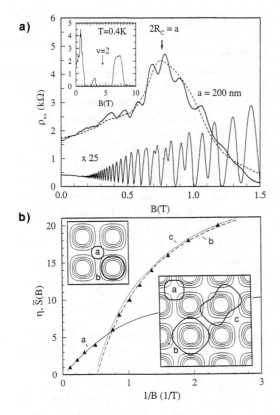

Fig. 1.2. (a) Diagonal resistivity ρ_{xx} measured in the patterned (*top curves*) and unpatterned (*bottom curve*) segment of the same sample for $T = 0.4$ K (*solid lines*) and 4.7 K (*dashed line*). At low B and $T = 0.4$ K, the quantum oscillations discovered for the antidot lattice are B-periodic, contrary to the usual Shubnikov–de Haas oscillations of the unpatterned sample, which scale with $1/B$. *Left inset*: $\rho_{xx}(B)$ for the antidot lattice up to 10 T. At high field, the emergence of Shubnikov–de Haas oscillations reflects the quantization of essentially unperturbed cyclotron orbits. (**b**) The *triangles* mark all $1/B$ positions of the ρ_{xx} minima. At high B the resistance minima lie equidistant on the $1/B$ scale; at low B the spacing becomes periodic in B. The *solid, dashed* and *dotted* lines are calculated reduced actions $\tilde{S}(1/B)$ of orbits (a), (b), and (c), respectively. These orbits are shown for $1/B = 0.6$ T^{-1} (*top*) and $1/B = 2.7$ T^{-1} (*bottom inset*). (From [51], by permission; ©1993 by the American Physical Society)

more detail in Sect. 3.1. Here we focus on measurements of the magneto-resistivity at low temperature.

In Fig. 1.2a, the diagonal resistivities, $\rho_{xx}(B)$, from both patterned and unpatterned segments of the same sample, are compared. The upper curves show $\rho_{xx}(B)$ for the antidot pattern. $\rho_{xx}(B)$ displays clear oscillations [51], superimposed upon a broad resistivity peak (dashed line) which persists up to high temperature and can be explained within a classical resistivity model. The temperature sensitivity of these oscillations – they are smeared out at 4.7 K – suggests that they are of a quantum nature. In the unpatterned part, $1/B$-periodic Shubnikov–de Haas (SdH) oscillations reflect the Landau energy spectrum of the two-dimensional bulk electron gas (bottom curve, shown on a magnified scale × 25). The oscillations in the antidot segment reveal a quite different behavior: they are B-periodic with a period corresponding to the addition of approximately one flux quantum through the antidot unit cell.

Assuming that ρ_{xx} reflects density-of-state oscillations,[10] the observed periodicity of these modulations could be semiclassically ascribed to quantized periodic orbits [50, 51] of electrons in the antidot landscape. An analysis on the basis of the Gutzwiller trace formula for the density of states, (2.27), shows that under the conditions of the antidot experiment (temperature and impurity broadening) only a few fundamental orbits play an essential role. They are depicted in the two insets of Fig. 1.2b at different magnetic field. The B dependence of the ρ_{xx} oscillations can be understood from the B dependence of the (quantized) classical actions of these periodic orbits,

$$S_{\mathrm{po}}(B) = \oint \left(m^* \boldsymbol{v} + \frac{e}{c} \boldsymbol{A} \right) \mathrm{d}\boldsymbol{r} = m^* \oint \boldsymbol{v} \mathrm{d}\boldsymbol{r} - \frac{e}{c} B \mathcal{A}_{\mathrm{po}}(B) . \qquad (1.1)$$

Here $B\mathcal{A}_{\mathrm{po}}(B)$ denotes the enclosed flux through a periodic orbit. For unperturbed cyclotron motion $\mathcal{A}_{\mathrm{po}}(B) = \pi R_{\mathrm{c}}^2$ scales with $1/B^2$, and $1/B$-periodic resistance oscillations result. In an antidot lattice the orbits cannot expand freely with decreasing field. To a first approximation the area enclosed by the orbits (b) and (c) shown in the lower right inset of Fig. 1.2b remains constant, $\mathcal{A}_{\mathrm{po}} \sim a^2$, causing the B-periodic oscillations with $B \approx h/ea^2$ displayed in Fig. 1.2a. The triangles in Fig. 1.2b, which mark all $1/B$ positions of the measured ρ_{xx} minima, lie exactly on curves representing the (reduced [51]) actions of the periodic orbits shown.

In Sect. 3.1 we shall give a quantitative description and a refined picture based on a semiclassical approach to the Kubo transport theory. There we shall see that the simple picture, invoked above, highlighting the role of a few periodic orbits, is essentially correct.

1.1.2 Ballistic Weak Localization in Electron Billiards

The second example deals with charge transport through phase-coherent quantum dots based on high-mobility semiconductor microstructures such as that displayed in Fig. 1.3. The resistance in a two-probe measurement for such a device is dominantly due to reflections of electrons at the walls of the confinement potential: the transport can be considered as ballistic. Depending on the size of the connections to the quantum dot, its spectrum ranges

---

[10] This may be inferred from the Shubnikov–de Haas oscillations of the pure 2DEG, which show a similar temperature dependence and mirror the Landau-level density of states at moderate fields [49].

from isolated resonances for small coupling to the exterior up to a regime of strongly overlapping resonant features, known as Ericson fluctuations in the context of nuclear physics [53].

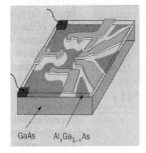
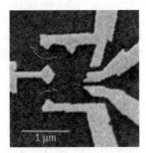

Fig. 1.3. A lateral quantum dot made by confining electrons at a two-dimensional interface between GaAs and AlGaAs by additional negatively charged surface gates (*light regions*). The *right figure* shows a micrograph of such an electron billiard where depletion around the gates (*light regions*) allows electrons to pass in and out of the dot only through the two left leads (from [54], by permission)

In an initial key experiment Marcus and Westervelt probed such Ericson fluctuations by measuring the related ballistic conductance fluctuations in electron billiards with the shape of a circle and of a stadium, the arena for chaotic dynamics [55].

Here, we focus on the weak-localization effect, a quantum enhancement of the average magnetoresistance at small magnetic field. As originally proposed by Baranger, Jalabert, and Stone [42,56], signatures of classical dynamics in electron billiards should be observable in quantum properties of the conductance, e.g. in weak-localization line shapes when displayed as a function of the magnetic field. Baranger et al. developed a semiclassical approach to the Landauer formalism, which connects the apparently macroscopic concept of conductance with microscopic scattering theory.

Semiclassically, the quantum transmission coefficient determining the current through a quantum dot can thereby be expressed through sums over pairs of phase-carrying trajectories. Constructive interference of an incoming path which is scattered from the cavity walls back to the entrance with its time-reversed partner contributes to the weak-localization enhancement of the resistance at zero magnetic field. A magnetic field leads to a dephasing of backscattered interfering waves traveling along time-reversed paths because the flux enclosed and hence the Aharonov-Bohm-type phases are of opposite sign.

The behavior of long multiply reflected trajectories and hence the mechanisms of accumulating flux differ for regular and generic chaotic geometries, as will be discussed in Sect. 3.2.3. Invoking such different classical dynamics, semiclassical transport theory predicts a *universal* Lorentzian line shape of the average resistance around $B = 0$ for geometries with pure chaotic classical dynamics. For integrable geometries the line shape is system-specific. For a circular billiard a linear decrease of the average resistance is expected.

In a further key experiment, Chang et al. [44] could indeed verify these predictions [56] by measuring the averaged resistance for ensembles of circular and stadium-type quantum dots. Their results are depicted in Fig. 1.4. The experiment of Chang et al. distinguishes between regular and chaotic quantum dots and shows in an impressive way the imprint of classical dynamics on a measured quantum effect.

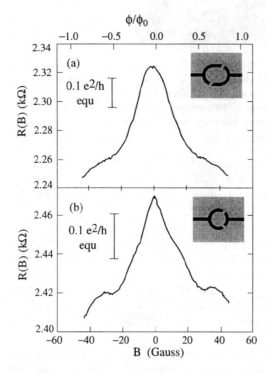

Fig. 1.4. Measured magnetoresistance of an ensemble of stadium-shaped (*top*) and circular (*bottom*) quantum dots. The different line shapes, Lorentzian for the stadium, linear for the circle, reflect the different classical electron dynamics. (From [44], by permission; ©1994 by the American Physical Society)

We see that semiconductor microstructures have become appropriate tools for the experimental study of scattering properties of integrable and chaotic systems. However, though semiclassics has provided the means to analyze and interpret the weak-localization line shapes, a full, quantitative semiclassical account of the quantum corrections to the conductance of ballistic systems is

still lacking (see Sect. 3.3). Ballistic weak localization remains as a paradigm
of modern semiclassical theory.

1.1.3 Mesoscopic Orbital Magnetism

Quantum phenomena are particularly spectacular in the case of orbital magnetism in the mesoscopic regime. On the one hand, according to the Bohr–
van Leeuwen theorem, there exists no classical orbital magnetism which could
mask quantum corrections. On the other hand, the Landau diamagnetism of
a bulk electron gas is tiny. Hence one can expect clear and pronounced mesoscopic quantum effects due to the confinement potential in quantum dots.
Nevertheless the orbital magnetic response of a quantum dot remains small
in magnitude and has challenged experimentalists to combine SQUID technology (to measure the magnetic behavior) with microstructure fabrication.
In this respect, the experiment by Lévy et al. [57], measuring the averaged
response of an ensemble of square quantum dots, can be regarded as a mile-
stone. Its main result, namely the enhanced magnetic susceptibility at small
magnetic field, is shown in Fig. 1.5. The experiment tells us that after par-
titioning a bulk 2DEG into an array of about 10^5 independent disconnected
squares, the orbital magnetism switches from the small *diamagnetic* Landau
bulk susceptibility to a *paramagnetic* response showing a drastic enhancement
by about a factor of 100!

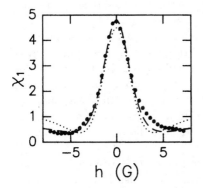

Fig. 1.5. Measured orbital magnetic susceptibility as a function of magnetic field
for an ensemble of $\sim 10^5$ square-shaped quantum dots with a mean size of 4 μm.
The *dotted* and *dashed* lines are empirical fits (from [57], by permission)

This experiment is comparable with the weak-localization measurements
above in so far as one detects a quantum mechanical enhancement of an
averaged quantity at zero magnetic field which decays on a magnetic-field
scale corresponding to approximately one flux quantum through the system.
While the weak localization can be understood in terms of long classical

trajectories, the enhanced magnetism for the integrable squares is mainly due to families of short flux-enclosing classical trajectories. However, an adequate theoretical understanding of the experimental observations requires not only a uniform semiclassical treatment of the dynamics in a billiard model, but also the inclusion of disorder *and* interaction effects. The corresponding semiclassical tools will be provided in Chaps. 4–6.

1.1.4 Andreev Billiards

As was illustrated above, measurements of the average conductance or magnetism can probe chaos in ballistic quantum devices. However, the mean density of states itself may also serve as an indicator of regular or irregular classical motion. As suggested by Beenakker and coworkers, this can be achieved by bringing a normal-conducting mesoscopic device like a semiconductor billiard into contact with a superconductor [58]. Such normal–superconducting hybrid structures have recently attracted attention as a new system in which quantum chaos can operate [25,59–62]. They were coined "Andreev billiards" [59] because the ballistic motion is modified owing to Andreev reflection at the interface with the superconductor. The proximity of the superconductor alters the density of quasiparticle states $d(E)$ of the normal conductor: it causes a depletion of excited states close above the Fermi energy known as the proximity effect. Beenakker et al. showed that a chaotic quantum dot, coupled to a superconductor via a ballistic point contact, exhibits a pronounced gap, whereas for an integrable rectangular structure the averaged density of states increases linearly without a gap above the Fermi energy.

This behavior is illustrated in Fig. 1.6. The left panel shows the mean density of states above the Fermi level, $E = 0$, for an Andreev billiard with rough boundaries (as sketched in the inset) which give rise to ergodic motion. The right panel shows the level density of a rectangular billiard with straight walls. The circles and full lines are results from numerical quantum mechanical calculations [62]. The difference in $d(E)$ for the two systems at small quasiparticle excitations is evident. It clearly allows one to distinguish chaotic and regular geometries.

Again semiclassics provides an explanation of this peculiar behavior. Semiclassically, the proximity effect arises from the unusual classical dynamics of paths which hit the interface between the superconductor and the quantum billiard. Such trajectories undergo specular reflection at the normal boundaries and Andreev reflection at the interface: the incoming electron-like quasiparticle is converted to a hole which is retroreflected and travels back along the path of the electron (see inset of Fig. 1.6). The density of states can be expressed in terms of distributions of path lengths $P(L)$ of Andreev-reflected orbits through a Bohr–Sommerfeld-type expression [58]

$$d(E) \simeq \frac{2w}{A\Delta} \int_0^\infty \mathrm{d}L \; P(L) L \sum_{n=0}^\infty \delta \left[\frac{EL}{\hbar v_{\mathrm{F}}} - \left(n + \frac{1}{2} \right) \pi \right] . \qquad (1.2)$$

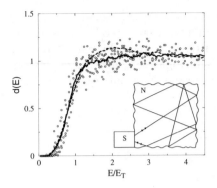
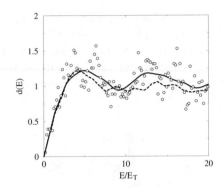

Fig. 1.6. Averaged density of states of a normal-conducting billiard (N) coupled to a superconductor (S in *inset*). The density of states is reduced close to the Fermi energy, $E = 0$, owing to the proximity effect. A chaotic Andreev billiard exhibits a pronounced gap (*left panel* for the case of a billiard with rough walls as sketched in the *inset*), while a rectangular billiard, representing integrable geometries, shows a linear increase of the density of states at the Fermi level (*right panel*). Quantum mechanical results (*circles*, and *solid lines* after averaging) are compared with a semiclassical prediction based on (1.2) (*dashed lines*). The energy is in units of $E_T = \hbar v_F w / 2\pi A$, where A is the billiard area and w the length of the interface with the superconductor. (Adapted from [62])

Here w is the length of the interface, A is the area of the billiard, and Δ the mean level spacing. Equation (1.2) holds for energies well below the excitation gap of the bulk superconductor, which is much larger than the energy scale $E_T = \hbar v_F w / 2\pi A$ in Fig. 1.6. Equation (1.2) can be derived either from the Eilenberger equation [61] or in the framework of a semiclassical S-matrix approach using the so-called diagonal approximation [62].

Different length distributions for regular and chaotic geometries lead to the evident differences in the depletion of $d(E)$. For the rough Andreev billiard with ergodic dynamics, $P(L)$ decreases exponentially, giving rise to an exponentially increasing $d(E)$ and the formation of a pseudogap. The rectangular billiard exhibits a power-law length distribution for large L. Hence the relatively large number of long orbits generates a linear increase in $d(E)$. The dashed curves in Fig. 1.6 show the corresponding semiclassical results, which agree fairly well with the averaged quantal values.

Summarizing, superconductor–semiconductor hybrid structures provide a new field for quantum chaos. These ideas open up the possibility, in principle, of measuring chaos in a small conductor by probing it from the outside with a superconducting tip.

1.1.5 Quantum Corrals

As the examples above have shown, information on quantum states in confined microdevices is usually gained only indirectly, e.g. by measuring mag-

netic or electric properties or statistical averages of these quantities; charge
densities of electronic wavefunctions are commonly not directly accessible. As
a last introductory example, where the latter is indeed achieved, we present
the quantum corrals of Crommie, Lutz, and Eigler [63], another fascinating
example of "quantum engineering" in mesoscopic electronic systems.

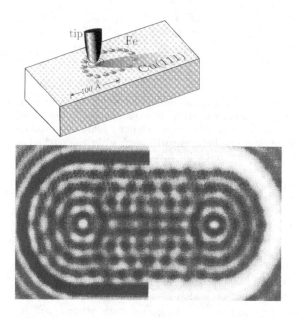

Fig. 1.7. *Upper panel*: an STM tip scanning the local density of states inside a
previously constructed quantum corral. *Lower panel*: comparison of theory (*left*)
and experiment (*right*) for a 76 atom stadium corral. (From [64], by permission)

Quantum corrals (Fig. 1.7) are made out of Fe adatoms which are indi-
vidually placed on a Cu(111) surface using the tip of a scanning tunneling
microscope (STM). The Fe atoms act as posts leading to scattering of (nearly
free) electrons at the Cu surface. When arranged like closed strings of pearls
the Fe adatoms can confine surface electronic states inside these geometries.
The STM tip then allows for the measurement of local densities of states,
$d(\boldsymbol{r}, E) = \sum_k |\psi_k(\boldsymbol{r})|^2 \delta(E - E_k)$, of the electronic surface waves $\psi_k(\boldsymbol{r})$. The
tip in this type of STM spectroscopy can also be viewed as a Green function
point source of electron amplitude. The induced amplitude is multiply scattered at the corral adatoms before returning to the tip. The backscattered
amplitude results in an interference term in the local density of states. The
bottom panel of Fig. 1.7 shows the impressive agreement of the measured surface wave pattern with the results from a parameter-free multiple-scattering
calculation for a stadium corral [64]. We see that these electronic objects have

inherently mesoscopic features: they rely on phase coherence and show inter-ference patterns with wavelengths much smaller than the confining device dimension.

Crommie, Lutz, and Eigler originally arranged the Fe adatoms in circular and stadium-shaped geometries with the intention of investigating features of wave chaos in surface electron waves, e.g. eigenfunction scars. Though the corrals turned out to be too leaky to produce quantum chaos, they instead opened up a variety of other theoretical questions: quantum corrals and re-lated surface structures, where targets can be arranged in a controlled way, have become a new arena for scattering by microscopic objects and scattering theory in two dimensions [65]. This direction, however, is beyond the scope of this book.

1.2 Purpose of This Book and Overview

The purpose of this book is both to present the general concepts of modern semiclassical theory and to put emphasis on their further development and application to problems in mesoscopic physics. It is our intention to demon-strate that semiclassical theory not only provides appropriate concepts to deal with problems of quantum chaos in general but also provides powerful tools to quantitatively compute mesoscopic quantities. To this end we review various applications of semiclassical approaches in mesoscopic systems and also compare the predictions with corresponding results from experiment or numerical quantum calculations.

All the quantities to be considered can be deduced from the single-particle Green function or its products. Hence its semiclassical representation in terms of a sum over contributions from classical paths provides the common basis of the work described here. The semiclassical evaluation of quantities such as spectral density, conductance, and magnetization, to name a few, then de-pends crucially on the character of the classical dynamics. It is on that level that the striking differences in the quantum signatures of integrable and chaotic systems enter. Typically, oscillatory quantum corrections are semi-classically represented as trace formulas, which represent physically trans-parent expressions of spectral or transport quantitites as sums over classical paths.

The present book is organized as follows. In the next chapter we present semiclassical techniques and expressions which serve as a general basis for the subsequent chapters, where they are adapted to particular problems. We briefly summarize trace formulas for the density of states and results for spectral correlations. We then introduce semiclassical approximations to thermodynamic quantities based on the spectral density and review a general semiclassical approach to dynamic linear response functions and dynamic susceptibilities.

Chapters 3 and 4 provide a semiclassical treatment of two prominent mesoscopic quantities: transport and orbital magnetism. Each chapter includes a brief but rather self-contained introduction to its respective area of mesoscopic physics, where the semiclassical methods are integrated into the general context.

In Chap. 3 we summarize the two complementary approaches to quantum transport. The semiclassical approximation to the Kubo conductivity yields, besides the classical part, quantum corrections in terms of a trace formula involving periodic orbits. In the semiclassical evaluation of the Landauer approach to conductance, interference between open classical orbits gives rise to quantum corrections. As an application of the two semiclassical approaches we discuss transport in antidot lattices [51, 66–68] and microcavities [33], and compare the results with corresponding experiments. Furthermore, we discuss the deficiencies which both approaches presently exhibit with regard to a correct treatment of weak-localization effects.

Geometrical effects of orbital magnetism in clean systems [45] are comprehensively reviewed in Chap. 4. There we discuss the magnetic susceptibility of singly connected geometries and the persistent current of ring-type geometries [69, 70]. We describe the fact that confinement does not essentially alter the small Landau diamagnetic response but gives rise to additional, strong quantum fluctuations in the magnetism of individual systems [71]. These fluctuations even survive as a quantum correction after ensemble averaging. In particular, these corrections are quite sensitive to the classical dynamics of the system. Hence we treat separately the cases of chaotic, integrable, and perturbed integrable classical behavior and compare them with numerical quantum calculations.

Throughout most of the book we address effects in *ballistic*[11] mesoscopic systems. For most of the applications, for instance conductivity, it then proves sufficient to treat particles in clean (disorder-free) systems and account for weak disorder effects by including an energy-independent broadening of the single-particle levels. This introduces an exponential damping of semiclassical path contributions on the scale of the elastic mean free path, but the major effects can still be related to the confinement potential.

The last two chapters are then devoted to a more thorough consideration of further mesoscopic effects due to disorder and electron–electron interaction. These sections are intended to give a more complete and realistic description of ballistic quantum systems and to evaluate the applicability and validity of the simple billiard models which are frequently used in studies of chaos in quantum mechanics.

Therefore, in Chap. 5 we present a refined semiclassical treatment of weak disorder [72, 73]. We study impurity effects on spectral correlation functions arising from the interplay between disorder and boundary scattering in the

---

[11] For reviews of semiclassical approaches devoted to *diffusive* systems we refer the reader to, for instance, [7, 76, 77].

whole regime from diffusive to ballistic. For ballistic quantum dots we put particular emphasis on the role of smooth disorder potentials. As an application, we readdress orbital magnetism and discuss how disorder affects the results from clean systems.

In Chap. 6, finally, we review semiclassical work on electron–electron interaction effects in mesoscopic systems. This approach allows one to reproduce results from quantum diagrammatic perturbation theory in the *diffusive* regime [74]. Moreover, a semiclassical perturbative treatment of interactions in the *ballistic* regime shows that the character of the classical dynamics of independent particles is reflected in the quantum properties of interacting ballistic systems [75].

If not explicitly stated otherwise, we assume throughout the parts of the work related to mesoscopic applications that the systems under consideration are phase-coherent; i.e. the phase-coherence length ℓ_ϕ is supposed to be larger than all other relevant length scales. Furthermore, we treat the electrons as spinless particles and include instead a degeneracy factor $g_s = 2$ whenever needed.

2. Elements of Modern Semiclassical Theory

The foundation of modern semiclassical theory is the fundamental work by Gutzwiller on the trace formula for the density of states of a chaotic quantum system. In a series of papers beginning in 1967 he built the bridge between the classical chaotic dynamics of a system and its quantum mechanical density of states [78–80]. His derivation is based on a semiclassical evaluation of the Feynman path integral in terms of saddle point approximations.

The trace formula has opened up a new field of physics: an overwhelming and still growing number of articles have dealt with various aspects of semiclassical physics in this general area. They span the whole range from more mathematically oriented investigations of convergence problems of semiclassical trace formulas to applications to various fields of physics and interpretations of numerical and experimental data.

For recent overviews of the subject, which cover at least parts of the literature, we refer the reader to the books by Gutzwiller [26], Ozorio de Almeida [27], Reichl [29], and Brack and Bhaduri [30] and, furthermore, to the collections of review articles in [12, 28]. A recent extensive overview by Gutzwiller of publications in the field of quantum chaos is found in [31].

The present chapter is mostly intended to provide some general semiclassical expressions which will be used as a platform for the subsequent chapters. There they will be further developed in the context of the specific mesoscopic applications addressed. Starting from the semiclassical Green function, we introduce the trace formulas for integrable and chaotic systems. We then give a brief summary of the success and shortcomings of semiclassical approaches to spectral correlation functions in Sect. 2.2. We present thermodynamic quantities in Sect. 2.3, which will serve as a basis of our approach to orbital magnetism. In Sect. 2.4 we introduce semiclassical techniques to evaluate linear response functions which involve the computation of matrix elements. These concepts will be applied to describe mesoscopic quantum transport.

2.1 Green Functions and Trace Formulas

Throughout most of this book we shall deal with time-independent systems in d dimensions described by a single-particle quantum Hamiltonian for spinless electrons of the form

$$\hat{\mathcal{H}} = \frac{1}{2m} \left(\hat{\boldsymbol{p}} - \frac{e}{c} \boldsymbol{A}(\hat{\boldsymbol{r}}) \right)^2 + V(\hat{\boldsymbol{r}}) \, . \tag{2.1}$$

$\boldsymbol{A}$ is the vector potential generating, for example, a magnetic field H. In the mesoscopic context, the one-body potential $V(\hat{\boldsymbol{r}})$ may, for instance, describe an electrostatic confinement potential, the mean field from other electrons, or a disorder potential.

The energy-dependent Green function $G(\boldsymbol{r}, \boldsymbol{r}'; E)$ provides a suitable means for investigating mesoscopic properties. It will be used to calculate transport quantities that are based on two-particle Green functions (products of single-particle Green functions for noninteracting systems), to deal with weak disorder and interaction effects, and to study those cases where a direct application of trace formulas is inappropriate, as will be discussed below. The retarded (or advanced) Green function can be expressed quantum mechanically as

$$G^{\pm}(\boldsymbol{r}, \boldsymbol{r}'; E) = \sum_{\lambda} \frac{\psi_\lambda(\boldsymbol{r}) \, \psi_\lambda^*(\boldsymbol{r}')}{E - E_\lambda \pm \mathrm{i}\epsilon} \tag{2.2}$$

in terms of the eigenfunctions ψ_λ and eigenenergies E_λ of the Hamiltonian $\hat{\mathcal{H}}$.

2.1.1 Semiclassical Green Function

Our basic starting point will be a semiclassical expression for the (retarded) energy-dependent single-particle Green function. It is of the form [26]

$$G^{\mathrm{sc}}(\boldsymbol{r}, \boldsymbol{r}'; E) = \sum_{t} G_t(\boldsymbol{r}, \boldsymbol{r}'; E)$$

$$= \frac{1}{\mathrm{i}\hbar (2\mathrm{i}\pi\hbar)^{(d-1)/2}} \sum_{t} D_t(\boldsymbol{r}, \boldsymbol{r}') \exp\left(\frac{\mathrm{i}}{\hbar} S_t - \mathrm{i}\eta_t \frac{\pi}{2} \right) . \tag{2.3}$$

It is given as a sum over contributions G_t from all classical trajectories t connecting the two fixed points $\boldsymbol{r}'$ and $\boldsymbol{r}$ at energy E.

$$S_t(\boldsymbol{r}, \boldsymbol{r}'; E) = \int_{\mathcal{C}_t} \boldsymbol{p} \cdot \mathrm{d}\boldsymbol{q} \tag{2.4}$$

is the classical action along a path $\mathcal{C}_t$ between $\boldsymbol{r}'$ and $\boldsymbol{r}$ and governs the accumulated phase. The semiclassical expression (2.3) can be derived from the Feynman path integral [93] within the stationary-phase approximation. G^{sc} is usually a good approximation to the exact Green function (2.2) provided the classical actions $S_t \gg \hbar$. This condition may be considered as defining the semiclassical regime.

The Maslov index η_t entering into the phase on the right-hand side of (2.3) is equal to the number of conjugate points on the trajectory [26,81]. Its calculation can be rather involved in practice.

The classical amplitude $D_t(r, r')$, describing the local density in position space and energy of trajectories near C_t, can be written as the determinant

$$D_t(r, r') = \begin{vmatrix} \dfrac{\partial^2 S_t}{\partial r \partial r'} & \dfrac{\partial^2 S_t}{\partial r \partial E} \\[2ex] \dfrac{\partial^2 S_t}{\partial E \partial r'} & \dfrac{\partial^2 S_t}{\partial E^2} \end{vmatrix}^{1/2} = \frac{1}{|\dot{q}_1 \dot{q}_1'|^{1/2}} \left| -\frac{\partial^2 S_t}{\partial q_i \partial q_j'} \right|^{1/2} . \qquad (2.5)$$

Here the indices $i = j = 1$ are excluded from the determinant on the right-hand side. Usually the q_i are chosen so as to span a local orthogonal coordinate system, with q_1 being a distance along and the other q_i being distances perpendicular to the trajectory t at point r, as introduced by Gutzwiller in his original derivation [79].[1] Then $\dot{q}_1$ denotes the velocity along the orbit at r.

The squares of the moduli of the classical amplitudes $D_t(r, r')$ can be linked to classical distribution functions in the configuration space: If we keep in the product $G^{\rm sc}[G^{\rm sc}]^* = \sum_{t', t} G_{t'} G_t^*$ only pairs of identical paths, their phases cancel and we are left with the classical contribution $\sum_{t=t'} |\tilde{D}_t|^2 \equiv 1/(2\pi\hbar^3) \sum_t |D_t|^2$ (in two dimensions). This can be expressed as

$$\sum_{t: r' \to r} |\tilde{D}_t|^2 = \frac{1}{2\pi\hbar^3} \int dq \, \delta(r - q) \int dE \, \delta[E - \mathcal{H}(p', r')] \begin{vmatrix} \dfrac{\partial p'}{\partial q} & \dfrac{\partial p'}{\partial E} \\[2ex] \dfrac{\partial \tau}{\partial q} & \dfrac{\partial \tau}{\partial E} \end{vmatrix} , \quad (2.6)$$

where we have used $\partial S/\partial E = \tau$, the classical traversal time, and $\partial S/\partial r' = -p'$ in D_t as given in (2.5). The determinant $|D_t|^2$ acts as a Jacobian for the transformation from E and q to to the new variables τ and p'. We then find, for the following related quantity,

$$\frac{\hbar}{\pi(\bar{d}/V)} \sum_{t: r' \to r} |\tilde{D}_t|^2 \delta(t - \tau_t)$$

$$= \frac{1}{\hbar^2 \pi(\bar{d}/V)} \int d\tau \int dp' \delta(t - \tau)\delta[r - q(\tau)] \, \delta[E - \mathcal{H}(p', r')] \qquad (2.7a)$$

$$= \frac{1}{\hbar^2 \pi(\bar{d}/V)} \int dp' \delta[r - q(t)]\delta[E - \mathcal{H}(p', r')]$$

$$= P(r', r; t) . \qquad (2.7b)$$

The right-hand side represents the classical density of trajectories at r after time t which begin at r' with arbitrary momentum p' on the energy shell. Hence (2.7b) relates the weights $|\tilde{D}_t|^2$ occurring in the Green function product to the classical probability $P(r', r; t)$ of propagating from r' to r in time t.

---

[1] Note, however, that the second equality in (2.5) holds in any coordinate system, as was shown by Littlejohn (see Sect. III.C in [82] and Sect. III in [83]), a property that sometimes can be employed to choose an appropriate local Poincaré surface of section.

The normalization of P contains the mean density of states $\bar{d}$, (2.23), per volume V (area in two dimensions).

All the classical entries D_t, S_t, and η_t in (2.3) usually have to be calculated numerically. For billiards, i.e. systems with free motion inside a confining geometry with a Dirichlet boundary condition, their computation is often simplified. Quantum billiards will play a major role throughout this work. First, they represent theoretically well-defined and thoroughly studied systems where, at least for some geometries, the classical dynamics can be proved to be completely chaotic. Despite their conceptual simplicity they generally exhibit complex classical and quantum behavior. Second, billiards constitute well-suited models for the ballistic microstructures which are experimentally realized. For billiards the magnitude p of the momentum is constant and it is convenient to introduce the wave number

$$k = \frac{p}{\hbar} = \frac{\sqrt{2mE}}{\hbar} . \tag{2.8}$$

The time of flight and the action integral (in the absence of a magnetic field) of a given trajectory can be simply expressed in terms of its length L_t as

$$\tau_t = \frac{m}{p} L_t ; \qquad \frac{S_t}{\hbar} = kL_t . \tag{2.9}$$

In billiard systems twice the number of reflections from the boundary must be added to η_t in order to take into account the phase π acquired at each bounce on the hard walls. We shall then still refer to η_t as the Maslov index, although slightly improperly.

The semiclassical Green function $G^{0,\mathrm{sc}}$ for a free particle in Euclidean space takes the form [26]

$$G^{0,\mathrm{sc}}(\boldsymbol{r}, \boldsymbol{r}'; E) = \frac{\pi}{E} \left(\frac{k}{2\pi \mathrm{i}} \right)^{(d+1)/2} \frac{\exp(\mathrm{i}k|\boldsymbol{r}' - \boldsymbol{r}|)}{|\boldsymbol{r}' - \boldsymbol{r}|^{(d-1)/2}} . \tag{2.10}$$

This equation is exact only for d odd. For $d=2$, the case on which we shall mainly focus throughout this book, it has the correct long-range asymptotic behavior of the exact free Green function $G^0(\boldsymbol{r}, \boldsymbol{r}'; E) = -\mathrm{i}m/(2\hbar^2) H_0^+(k|\boldsymbol{r}' - \boldsymbol{r}|)$, where H_0^+ denotes the standard Hankel function. However, the semiclassical amplitude in two dimensions,

$$D_t = \frac{m}{\sqrt{\hbar k |\boldsymbol{r} - \boldsymbol{r}'|}} , \tag{2.11}$$

does not show the correct logarithmic singularity of H_0^+ for $\boldsymbol{r}' \to \boldsymbol{r}$. Hence refined techniques, presented below, are required to treat this limit.

One main advantage of semiclassical concepts in mesoscopic physics is related to the fact that they allow us to account very efficiently for (small) changes in the original Hamiltonian. Frequently their effect can be considered to be negligible with respect to classical quantities but may be significant on quantum scales. This opens up the possibility to treat changes in (potentials of) the Hamiltonian semiclassically as a perturbation although they may be far beyond the range of applicability of quantum perturbation theory. The idea of this semiclassical perturbative approach is to keep the *unperturbed* paths $\mathcal{C}_t^0$ and classical amplitudes $D_t^0(E)$ in the Green function and consider merely the effect of the perturbation on the classical actions, as a variation δS_t;

$$S_t = S_t^0 + \delta S_t . \tag{2.12}$$

The related phase shift $\delta S_t/\hbar$ which – by means of the large factor $1/\hbar$ – may be considerable, accounts for the changes on quantum scales. Later in this book this approach will be applied in the following situations:

(i) For the case of a weak potential $\delta V(\boldsymbol{r})$ affecting free motion at energy E in a billiard, the correction term δS_t is given, after expanding $p = \sqrt{2m[E - \delta V(\boldsymbol{r})]}$ for small $\delta V/E$, by the integral

$$\delta S_t = -\frac{1}{v_{\mathrm{F}}} \int_{\mathcal{C}_t^0} \delta V(\boldsymbol{q})\, \mathrm{d}q . \tag{2.13}$$

This relation will be used in Chap. 5 to include the effect of a weak disorder potential.

(ii) For closed orbits, which are relevant in Green function trace formulas, one can employ a general result from classical mechanics [27, 84]. This is that the change in the action integral along a closed orbit at constant energy under the effect of a parameter λ of the classical Hamiltonian $\mathcal{H}$ is given by

$$\left(\frac{\partial S}{\partial \lambda}\right)_E = -\oint \mathrm{d}t\, \frac{\partial \mathcal{H}}{\partial \lambda} , \tag{2.14}$$

where the integral is taken along the unperturbed trajectory. As an important case, to be used frequently in the remainder of this book, we treat the effect of a small uniform static magnetic field H, which enters by means of the vector potential $\boldsymbol{A}(\hat{\boldsymbol{r}})$ into the Hamiltonian (2.1). Classical perturbation theory then yields

$$\delta S_t = \frac{e}{c}\, H \mathcal{A}_t^0 , \tag{2.15}$$

where $\mathcal{A}_t^0$ is the directed area enclosed by the unperturbed orbit.

(iii) Similar expansions are usually applied to account for small changes in energy, namely

$$S_t(E + \epsilon) \simeq S_t^0(E) + \frac{\partial S_t^0}{\partial E}\epsilon = S_t^0(E) + \tau_t^0\, \epsilon , \tag{2.16}$$

where $\tau_t^0(E) \equiv \mathrm{d}S_t^0/\mathrm{d}E$ is the period of the unperturbed orbit.

2.1.2 Density of States

The quantum mechanical density of states

$$d(E) = g_{\rm s} \sum_\lambda \delta(E - E_\lambda) \,, \tag{2.17}$$

where $g_{\rm s} = 2$ denotes the spin degeneracy and E_λ the eigenenergies, is related to the trace of the energy-dependent Green function $G(r, r'; E)$ by

$$d(E) = -\frac{g_{\rm s}}{\pi} \,\text{Im}\, \mathcal{G}(E) \quad ; \quad \mathcal{G}(E) = \int dr\, G(r, r; E) \,. \tag{2.18}$$

A semiclassical treatment of the Green function leads to a natural representation of the density of states as

$$d(E) = \bar{d}(E) + d^{\rm osc}(E) \,. \tag{2.19}$$

This decomposition into a smooth part (denoted by an overbar) and an oscillating part (denoted by the superscript "osc") has a rigorous meaning only in the semiclassical regime ($E \to \infty$), for which the scales of variation of $\bar{d}$ and $d^{\rm osc}$ decouple. We shall see that the decomposition into smooth and oscillating terms is a typical feature of spectral quantities in the mesoscopic regime.

The smooth part, $\bar{d}(E)$, is connected to contributions to the trace from the Green function in the limit $r \to r'$. Since the singularity of $G(r, r'; E)$ for $r \to r'$ is logarithmic in two dimensions, it is not appropriate to use the semiclassical Green function (2.10) in this limit. Instead it is convenient to employ the Wigner transform of the Green function to compute the contribution to the trace of direct paths for $r \to r'$. The Wigner transform of an operator $\hat{A}$ is defined as [27, 85]

$$A_{\rm W}(r, p) \equiv \int dx\, e^{-ipx/\hbar} \left\langle r + \frac{x}{2} \left| \hat{A} \right| r - \frac{x}{2} \right\rangle \,. \tag{2.20}$$

To leading order in $\hbar$ the Wigner transform of the Green function is [86]

$$G_{\rm W}(r, p, E) \simeq \frac{1}{E - \mathcal{H}(r, p)} \,. \tag{2.21}$$

In this form it accounts for "paths of zero length". Using the relation

$$\text{Tr}\, \hat{A} = \frac{1}{(2\pi\hbar)^d} \int dr\, dp\, A_{\rm W}(r, p) \tag{2.22}$$

for the trace of the Green function, we find the familiar Weyl or Thomas–Fermi part of the density of states:

$$\bar{d}(E) = \frac{g_{\rm s}}{(2\pi\hbar)^d} \int dr dp\, \delta\left[E - \mathcal{H}(r, p) \right] \,. \tag{2.23}$$

This is the leading-order semiclassical expression for the smoothed part of the density of states[2] and may be regarded as stemming from orbits of "zero length". The Weyl term reflects the volume of accessible classical phase space at energy E (divided by the quantum phase space cell).

Oscillatory terms in the density of states arise from contributions to $G(\boldsymbol{r}, \boldsymbol{r}; E)$ of paths of finite length closed in position space ($\boldsymbol{r} = \boldsymbol{r}'$), which we shall refer to in the following as *recurrent* orbits. The standard route for obtaining d^{osc} is to evaluate the trace integral (2.18) by the stationary-phase approximation. This selects those trajectories which are not only closed in configuration space but also closed in phase space ($\boldsymbol{r}' = \boldsymbol{r}, \boldsymbol{p}' = \boldsymbol{p}$), i.e. *periodic* orbits. If they are well isolated the Gutzwiller trace formula [79] for the density of states is obtained. [3] For integrable systems all recurrent orbits are in fact periodic since the action variables are constants of motion. Hence, periodic orbits in integrable systems are organized in continuous families associated with resonant tori. In this case a proper evaluation of the trace integrals leads to the Berry–Tabor [88, 89] or Balian–Bloch [90] trace formula. We first present these trace formulas, which represent the two opposite limiting cases of "hard chaos" and integrable dynamics, before discussing their limitations.

2.1.3 Berry–Tabor Trace Formula

We discuss in some detail the Berry–Tabor trace formula for integrable systems since it will be used to derive magnetic properties of integrable microstructures in Sect. 4.6. Furthermore, it helps one to understand the structure of a similar trace formula for matrix elements to be employed in Sect. 2.4 in the semiclassical approach to linear response functions. Finally, we shall consider in Sect. 4.4 modifications to the Berry–Tabor formula when the integrability is broken owing to a perturbation.

A classical Hamiltonian $\mathcal{H}(\boldsymbol{r}, \boldsymbol{p})$ is integrable if the number of constants of motion in involution is equal to the number of degrees of freedom [26]. For bounded systems, this implies that all trajectories are confined on invariant tori. Each torus is labeled by the action integrals [91]

$$I_i = \frac{1}{2\pi} \oint_{\mathcal{C}_i} \boldsymbol{p} \, \mathrm{d}\boldsymbol{r} \, . \tag{2.24}$$

For systems with two degrees of freedom the integrals are taken along two independent paths $\mathcal{C}_1$ and $\mathcal{C}_2$ on the torus. Then it is convenient to perform a canonical transformation from the original $(\boldsymbol{p}, \boldsymbol{r})$ variables to the action–angle variables $(\boldsymbol{I}, \boldsymbol{\varphi})$, where $\boldsymbol{I} = (I_1, I_2)$ and $\boldsymbol{\varphi} = (\varphi_1, \varphi_2)$ with φ_1, φ_2 in $[0, 2\pi]$. Because I_1 and I_2 are constants of motion, the Hamiltonian expressed in

---

[2] See also Sect. 4.2.2, where higher-order $\hbar$ corrections are considered. For general methods to calculate higher-order terms see [87].

[3] See also [90] for the derivation of a related semiclassical trace formula.

action–angle variables depends only on the actions, i.e. $\mathcal{H}(I_1, I_2)$. We call $\nu_i = \partial\mathcal{H}/\partial I_i$ $(i = 1, 2)$ the angular frequencies of the torus and $\alpha \equiv \nu_1/\nu_2$ the winding number. A torus is called "resonant" if the winding number is rational, i.e. $\alpha = u_1/u_2$, where u_1 and u_2 are coprime integers. The orbits on a resonant torus are periodic. Hence, the torus constitutes a one-parameter family of periodic orbits, each member of the family having the same period and action. The families of periodic orbits can be labeled by two integers $(M_1, M_2) = (ju_1, ju_2)$, where (u_1, u_2) specifies the primitive orbits and j is the number of repetitions. The pair $\boldsymbol{M} = (M_1, M_2)$ represents the "topology" of the orbits.

For two-dimensional systems, the Berry–Tabor trace formula for the oscillating part of the density of states can be written as [88, 89]

$$d^{\text{osc}}(E) = \sum_{\boldsymbol{M}, \epsilon} d_{\boldsymbol{M}, \epsilon}(E) \tag{2.25}$$

$$= \frac{g_s}{\pi \hbar^{3/2}} \sum_{\boldsymbol{M} \neq (0,0), \epsilon} \frac{\tau_{\boldsymbol{M}}}{M_2^{3/2} \left| g_E''(I_1^{\boldsymbol{M}}) \right|^{1/2}} \cos\left(\frac{S_{\boldsymbol{M}, \epsilon}}{\hbar} - \eta_{\boldsymbol{M}} \frac{\pi}{2} + \gamma \frac{\pi}{4} \right).$$

In (2.25), the sum includes all families of closed orbits at energy E. They are labeled by $\boldsymbol{M}$, where M_1 and M_2 are positive. Except for self-retracing orbits, the additional index ϵ specifies tori of the same topology, but related to each other through time reversal symmetry. $S_{\boldsymbol{M}, \epsilon}$ and $\tau_{\boldsymbol{M}}$ are the action integral and the period along the periodic orbits of the family $\boldsymbol{M}$, and $\eta_{\boldsymbol{M}}$ denotes the Maslov index.

The energy surface E in action space, which is implicitly defined through $\mathcal{H}(I_1, I_2) = E$, is explicitly given by the function $I_2 = g_E(I_1)$. The action variables of the resonant torus with periodic orbits of topology $\boldsymbol{M}$ are denoted by $\boldsymbol{I}^{\boldsymbol{M}} = (I_1^{\boldsymbol{M}}, I_2^{\boldsymbol{M}})$. They are determined by the condition

$$\alpha = -\left. \frac{\mathrm{d} g_E(I_1)}{\mathrm{d} I_1} \right|_{I_1 = I_1^{\boldsymbol{M}}} = \frac{M_1}{M_2}. \tag{2.26}$$

The first equality arises from the differentiation of $\mathcal{H}[I_1, g_E(I_1)] = E$ with respect to I_1. The last contribution to the phase in (2.25) is given by $\gamma = \text{sgn}[g_E''(I_1^{\boldsymbol{M}})]$.

The first derivation of the Berry–Tabor trace formula [88] starts from an EBK (Einstein, Brillouin, and Keller) quantization condition followed by the application of the Poisson summation rule (4.4). It involves a stationary-phase approximation valid in the semiclassical limit $S \gg \hbar$ with a stationary-phase condition according to (2.26).

2.1.4 Gutzwiller Trace Formula

The density of states of a classically chaotic quantum system, where all periodic orbits are unstable and well isolated, can be semiclassically described in terms of Gutzwiller's celebrated trace formula. The oscillating part of the density of states can be approximated to leading order in $\hbar$ as [26, 79]

$$d^{\text{osc}}(E) \simeq \frac{g_s}{\pi\hbar} \sum_{\text{po}} \sum_{j=1}^{\infty} \frac{\tau_{\text{po}}}{\left|\det(M_{\text{po}}^j - I)\right|^{1/2}} \cos\left[j\left(\frac{S_{\text{po}}}{\hbar} - \eta_{\text{po}}\frac{\pi}{2}\right)\right]. \quad (2.27)$$

The double sum is taken over contributions from all classical primitive periodic orbits labeled "po", and j denotes their multiple traversals. S_{po} is the action (2.4) and τ_{po} the period of a prime periodic orbit. The monodromy matrix M_{po} takes into account in a linearized way the phase space characteristics on the energy shell close to each periodic trajectory. M_{po} characterizes the instability of an orbit in the chaotic case measured in terms of Liapunov exponents. The Maslov index η_{po} equals twice the number of complete rotations of the eigenvectors of M_{po} around a tangent vector of the orbit plus twice the number of bounces off the walls for billiards. For systems with time reversal symmetry the nonself-retracing orbits have to be counted twice.

The trace formula provides the basic connection between a pure quantum property, namely spectral-density oscillations, and pure classical entities of the corresponding chaotic dynamic system. Interference effects occur via the semiclassical phases determined by the classical actions. The close similarity of the trace formula to a Fourier series suggests using it as a basis for the spectral analysis of complex spectra. In billiard systems, for example, the actions scale linearly with wave vector, $S_{\text{po}}/\hbar = kL_{\text{po}}$. Hence a Fourier transform with respect to k reveals peaks in the power spectrum of the density of states at positions marked by the lengths L_{po} of the periodic orbits (for an example see Sect. 3.2.2, Fig. 3.7). Furthermore, the trace formula implies the composition of a quantum spectrum in a hierarchical manner: short periodic trajectories produce long-ranged cosine-like spectral fluctuations. The longer the orbits included are, the higher is the spectral resolution of a semiclassically approximated quantum spectrum. Frequently, in particular in mesoscopics, experimentally obtained spectra are smeared and of limited resolution. Then the hierarchical structure of semiclassical trace formulas provides a means to extract the main spectral information on the basis of a limited number of (shorter) periodic orbits without performing a complete (quantum) computation of individual eigenstates of the system.

The literature on the trace formula (2.27) and its derivations has become so abundant that we only briefly present and discuss it here. We refer the reader to existing reviews [26–30, 34, 36, 92] and references therein for further aspects.

The semiclassical computation of the trace integral (2.18) leading to the trace formula (2.27) can be regarded as a special case of the trace integral (A.20) (with $A(r') \equiv 1$, $\langle A_t^{++} \rangle = \tau_t$), which is calculated in Appendix A.1.

For an early detailed review of the derivation of the trace formula in the framework of Feynman's path integral see, for example, [93]. In addition, several alternative derivations have been proposed, of which we list a few below.

In the work of the "Copenhagen school" of Cvitanović [34] the Gutzwiller trace formula is embedded in a more general framework of dynamical zeta functions and can be regarded as a semiclassical version of the latter. The zeta function formalism provides an appropriate tool to account for the properties of the underlying classical evolution in phase space. "Cycle expansions" allow for a reordering of the entries in the trace formula in order to improve its convergence [34, 35].

Bogomolny [92] reduces the semiclassical solution of the Schrödinger equation to the transfer operator T, which acts as a semiclassical Poincaré map on a subspace of the full phase space. T is constructed from semiclassical Green functions containing (short) paths from one iteration of the Poincaré map. The zeros of the related Fredholm determinant $\det(1-T)$ coincide with the poles of the Gutzwiller trace formula if the determinant is evaluated semiclassically within the stationary-phase approximation. The approach of Bogomolny has the advantage that it also in principle allows one to compute "semiquantum" energies: in such a computation the semiclassical T is used, but the Fredholm determinant is (numerically) calculated without further approximation (see, e.g., [94]). Furthermore, the combination of Fredholm theory and Bogomolny's approach provides a mathematically profound scheme [95] for the ordering of periodic-orbit contributions in trace formulas of chaotic systems.

The scattering formalism developed by Smilansky and coworkers [96] represents an alternative approach to the semiclassical spectral density. The density of states of a billiard is expressed in terms of traces of related S-matrices, which are evaluated semiclassically.

We note that there exists, furthermore, an elaborate time-dependent semiclassical treatment by Heller and Tomsovic [97]. By evaluating semiclassical propagators in the time domain these authors avoid complications which may arise owing to the additional semiclassical approximation involved in the transformation to the energy Green function.

Three major issues arise in connection with the trace formula and have received considerable attention in the literature over the last decade:

(i) Despite numerous approaches (e.g. [98–101]) the theory of the computation of semiclassical trace integrals is still incomplete for systems with nonuniform, in particular mixed, phase space. This shortcoming is severe since such systems represent the majority of nonintegrable systems. The trace formulas cited above for integrable systems and "hard chaos" mark only the two extreme limits of the broad spectrum of classical Hamiltonian dynamics. The "semiquantum" version of Bogomolny's approach is not restricted to either of these limits and can in principle deal with mixed-phase-space dynamics; however, it then remains on the numerical level. In the present work we mainly rely on the two limiting cases and enter into the intermediate regime in a perturbative approach starting from the regular case.

(ii) The lack of convergence of trace formulas and Green function expressions for chaotic systems at real values of energy. This divergence is related to the property that the number N_{po} of periodic trajectories in ergodic systems (more precisely axiom A systems) increases exponentially with period, i.e. $N_{\text{po}}(\tau_{\text{po}} < \tau) \simeq \exp(h\tau)/h\tau$, where h is the topological entropy. The exponential proliferation of the orbits is not sufficiently compensated by the corresponding decrease in the classical amplitudes. This issue has become a research direction of its own in quantum chaos. Various techniques have been employed to obtain appropriate analytical continuations of the trace formula [33, 102]. We shall be not concerned with such convergence problems, since, in typical mesoscopic applications, mechanisms such as temperature smoothing, residual impurity scattering, and inelastic processes usually lead to an additional natural damping of periodic-orbit contributions as will be discussed. Though there exist several examples of hyperbolic systems where individual energy levels have been obtained semiclassically with reasonable accuracy [33,35,101], the general problem of computing the quantum density of states with a resolution beyond the mean level spacing persists.

(iii) Related to this problem is the long standing issue of a semiclassical computation of spectral correlation functions down to scales of the mean level spacing. This will be discussed in the next section.

2.2 Spectral Correlations

Spectral correlation functions are important measures for describing density-of-states properties of mesoscopic quantum systems. Correlation functions are particularly helpful if one is interested in the statistical properties of energy level spectra instead of computing individual energy levels. One prominent tool is the two-point level-density correlator, which can be defined as

$$K(\epsilon) = \frac{1}{\bar{d}^2} \left\langle d\left(E + \frac{\epsilon}{2}\right) d\left(E - \frac{\epsilon}{2}\right) \right\rangle . \tag{2.28}$$

The angle brackets stand for ensemble averaging in the case of disordered systems or an average over an appropriate energy interval for a single chaotic system. In this book we are mainly concerned with computing $K(\epsilon)$ and related correlation functions for energy scales $\epsilon > \Delta$. The question of how to treat this correlation function in a proper semiclassical way in the entire ϵ range is not in the main focus of this book. However, since this question represents one of the main issues of the present studies of quantum chaos[4] we briefly make reference to some of the recent semiclassical developments in

---

[4] For a recent collection of articles addressing this issue see [103].

this area. Moreover, it enables us to introduce a few basic notions of averaged spectral quantities.

The above correlation function has been studied by Altshuler and Shklovskii in their seminal work [105] for noninteracting electrons in a *disordered* system using diagrammatic perturbation theory. These authors found for $K(\epsilon)$, besides the trivial mean part, the perturbative term

$$K^{\mathrm{P}}(\epsilon) = -\frac{1}{2\alpha\pi^2}\frac{\partial^2}{\partial\epsilon^2}\ln[D(\epsilon)]\ . \tag{2.29}$$

$D(\epsilon)$ is the spectral determinant of the (classical) diffusion operator. In (2.29) $\alpha = 2$ for non-time-reversal-symmetric systems (corresponding to a random-matrix ensemble denoted GUE) and $\alpha = 1$ for time-reversal-symmetric systems (GOE).

Later, Argaman, Imry, and Smilansky rederived, extended, and interpreted the diagrammatic results by using a semiclassical method for the diffusive regime [76] (see Sect. 5.5.1, Fig. 5.4). More recently, Andreev and Altshuler [106] computed the nonperturbative, oscillatory "correction". Intriguingly, it is also expressed in terms of $D(\epsilon)$. For example, in the unitary case, it reads

$$K^{\mathrm{osc}}(\epsilon) = -\frac{\cos(2\pi\bar{d}\epsilon)}{2\pi^2}D(\epsilon)\ . \tag{2.30}$$

A semiclassical treatment of the level correlator K for a chaotic system was originally performed by Berry in his seminal paper [107] on the Fourier transform of $K(\epsilon)$, the form factor

$$\tilde{K}(\tau) = \bar{d}\int_{-\infty}^{\infty}[K(\epsilon) - 1]\exp(2\pi\mathrm{i}\,\bar{d}\epsilon\tau)\,\mathrm{d}\epsilon\ . \tag{2.31}$$

The standard semiclassical analysis of K and $\tilde{K}$ is based on the so-called *diagonal approximation*. It involves considering only pairs of the same orbits and pairs of symmetry-related orbits in the double sum of periodic orbits which appears when using the Gutzwiller trace formula (2.27) for $d(E)$ in (2.28). Their dynamical phases $S_{\mathrm{po}}(E\pm\epsilon/2)/\hbar \simeq (S_{\mathrm{po}}(E)\pm\tau_{\mathrm{po}}\epsilon/2)/\hbar$ nearly cancel (up to $\tau_{\mathrm{po}}/\hbar$) so that these contributions persist on (energy) average. The corresponding diagonal contribution to $K(\epsilon)$ then reads

$$K^{\mathrm{diag}}(\epsilon) = \frac{2}{T_{\mathrm{H}}^2}\sum_{\mathrm{po}}\sum_{j=1}^{\infty}\frac{\tau_{\mathrm{po}}^2}{|\det(\boldsymbol{M}_{\mathrm{po}}^j - \boldsymbol{I})|}\cos\left(\frac{j\tau_{\mathrm{po}}\epsilon}{\hbar}\right)\ , \tag{2.32}$$

where

$$T_{\mathrm{H}} = 2\pi\hbar\bar{d} \tag{2.33}$$

denotes the Heisenberg time. Off-diagonal pairs of orbits which differ in their actions and related phases are often assumed to average out. This is justified if the actions of different orbits are not correlated and the off-diagonal contribution can be regarded as a sum of random phases.

The remaining single sum (2.32) is usually further evaluated using the Hannay–Ozorio de Almeida sum rule [27, 108]

$$\sum_{\text{po}} \frac{A_{\text{po}}}{|\det(\boldsymbol{M}_{\text{po}} - \boldsymbol{1})|} \simeq \int d\tau \ A(\tau)/\tau \ . \tag{2.34}$$

The evaluation of K or $\tilde{K}$ in the diagonal approximation using the above sum rule requires long times τ such that the orbits explore the energy shell uniformly. One then finds

$$K^{\text{diag}}(\epsilon) \simeq -\frac{1}{\alpha \pi^2 \bar{d}^2 \epsilon^2} \ . \tag{2.35}$$

Here the difference in α between systems without and with time-reversal symmetry stems from the existence of pairs of time-reversed orbits in the latter case.[5] The corresponding diagonal term of the form factor increases linearly with time: $\tilde{K}^{\text{diag}}(\tau) \sim \tau/\alpha$. The form factor coincides with the random-matrix GUE form factor for times smaller than T_{H}. However, $\tilde{K}^{\text{diag}}$ deviates strongly from the GUE result for $\tau \geq T_{\text{H}}$ (see e.g. Fig. 5.4). This fact implies that contributions from off-diaogonal terms must be considered.

Agam, Altshuler, and Andreev [46] showed that, for chaotic systems, the diagonal part $K^{\text{diag}}(\epsilon)$ of $K(\epsilon)$ is of the same functional form (2.29) as K^{P} (for diffusive systems), but with the spectral determinant $D(\epsilon)$ associated with the squared modulus of the dynamical zeta function [34] of the chaotic system. Correspondingly, they identified the diffusion operator of disordered systems with the Perron–Frobenius operator [34] of clean chaotic systems and thereby achieved a link between these two classes of ergodic systems. More recently, this identification was confirmed by a field-theoretical approach to chaotic systems [48]. See also [47] for a related approach.

Bogomolny and Keating [104, 111] proposed a novel semiclassical method to compute off-diagonal terms by relating them to the diagonal contributions. In this way they derived results close to those of [46]. At present, there exists an ongoing intense theoretical activity which aims at a better semiclassical understanding of nondiagonal contributions, a proper description of the long-time limit, the inclusion of diffraction effects, and the study of possible correlations between periodic orbits (e.g. [109, 110]). A final answer to this outstanding problem in quantum chaos is still missing.[6] Semiclassical methods remain very promising tools to resolve the question of how the classical limit of quantum systems, which is well described by random-matrix theory, is approached and how to incorporate system-specific corrections beyond random-matrix theory.

---

[5] For a more careful analysis of the role of such action multiplicities in this context see [104].

[6] See also [103] for recent work on the relation between field-theoretical and periodic-orbit approaches to spectral correlations.

2.3 Thermodynamic Quantities

One main subject of the present book is the introduction of semiclassical concepts into the thermodynamics of mesoscopic systems. Finite temperature leads to a smoothing of spectral quantities and yields a natural cutoff in semiclassical trace formulas due to a damping of contributions from long trajectories on the scale of a thermal cutoff length L_T.

Here, following [45], we deduce basic expressions for thermodynamic quantities, e.g. the temperature-smoothed spectral density, the particle number function, and the grand potential, from the quasiclassically calculated single-particle density of states of noninteracting particles. These expressions will serve as a convenient starting point for calculating some of the transport properties in Chap. 3, as well as magnetic quantities in Chap. 4. To avoid repetition, we shall consider here in some detail the thermodynamic properties and shall refer to the results obtained in this section whenever needed.

In the following we denote the chemical potential or Fermi energy by μ. The corresponding Fermi velocity and wave vector are $v_F = v(\mu)$ and $k_F = mv_F/\hbar$, respectively. Starting from the quantum mechanical single-particle density of states $d(E)$, (2.17), we define its related energy integrals, namely the energy staircase function and the grand potential, at zero temperature by

$$n(E) = \int_0^E dE'\, d(E') \quad ; \quad \omega(E) = -\int_0^E dE'\, n(E') \ . \qquad (2.36)$$

At finite temperature the corresponding quantities (denoted by capital letters) represent the number of particles in the grand canonical ensemble,

$$N(\mu) = \int_0^\infty dE\, d(E)\, f(E-\mu) \ , \qquad (2.37)$$

and the grand thermodynamic potential

$$\Omega(T,\mu,H) = -\frac{1}{\beta} \int dE\, d(E)\, \ln\{1 + \exp[\beta(\mu-E)]\} \ . \qquad (2.38)$$

Here, $\beta = 1/k_B T$ and

$$f(E - \mu) = \frac{1}{1 + \exp[\beta(E - \mu)]} \qquad (2.39)$$

is the Fermi distribution function. Integration by parts leads to the equivalent expressions

$$D(\mu) = -\int_0^\infty dE\, d(E)\, f'(E-\mu) \ , \qquad (2.40a)$$

$$N(\mu) = -\int_0^\infty dE\, n(E)\, f'(E-\mu) \ , \qquad (2.40b)$$

$$\Omega(\mu) = -\int_0^\infty dE\, \omega(E)\, f'(E-\mu) \ , \qquad (2.40c)$$

in terms of convolutions with the derivative of the Fermi function. This is a convenient representation for a semiclassical treatment since the derivative f' contributes predominantly at energies μ where semiclassical approximations are reliable.

According to the semiclassical decomposition (2.19) of the density of states, the above thermodynamic expressions can also be separated in a natural way into smooth and oscillating parts. At this point we need not distinguish between the representations of d^{osc} for integrable dynamics (Berry–Tabor trace formula (2.25)) and chaotic dynamics (Gutzwiller trace formula (2.27)) and we write

$$d^{\mathrm{osc}}(E) = \sum_{\mathrm{po}} d_{\mathrm{po}}(E) \;,$$

$$d_{\mathrm{po}}(E) = A_{\mathrm{po}}(E) \sin\left(\frac{S_{\mathrm{po}}(E)}{\hbar} + \nu_{\mathrm{po}}\right) , \qquad (2.41)$$

where "po" labels isolated periodic orbits and their multiples in the chaotic case as well as periodic trajectories representing tori in the integrable case.[7]

Using the expression (2.41) for d^{osc} in (2.36) one obtains the corresponding quantities

$$n^{\mathrm{osc}}(E) = \int^{E} \mathrm{d}E'\, d^{\mathrm{osc}}(E') \quad ; \quad \omega^{\mathrm{osc}}(E) = -\int^{E} \mathrm{d}E'\, n^{\mathrm{osc}}(E') \,. \,(2.42)$$

In a leading-$\hbar$ calculation the energy integrations in (2.42) have to be applied only to the rapidly oscillating phase of each periodic-orbit contribution $d_{\mathrm{po}}(E)$ and not to the amplitude $A_{\mathrm{po}}(E)$, which usually varies smoothly with energy. Employing, moreover, $\tau_{\mathrm{po}}(E) \equiv \mathrm{d}S_{\mathrm{po}}/\mathrm{d}E$, one has

$$\int^{E} A_{\mathrm{po}}(E') \sin\left(\frac{S_{\mathrm{po}}(E')}{\hbar} + \nu_{\mathrm{po}}\right) \mathrm{d}E'$$

$$= -\frac{\hbar A_{\mathrm{po}}(E)}{\tau_{\mathrm{po}}(E)} \cos\left(\frac{S_{\mathrm{po}}(E)}{\hbar} + \nu_{\mathrm{po}}\right)$$

$$\equiv -\frac{\hbar}{\tau_{\mathrm{po}}(E)} \tilde{d}_{\mathrm{po}}(E) \,. \qquad (2.43)$$

The integration over energy merely amounts to a multiplication by $(-\hbar/\tau_{\mathrm{po}})$ and a phase shift of $\pi/2$. We then get

$$n^{\mathrm{osc}}(E) = \sum_{\mathrm{po}} n_{\mathrm{po}}(E) \quad ; \quad n_{\mathrm{po}}(E) = -\frac{\hbar}{\tau_{\mathrm{po}}(E)} \tilde{d}_{\mathrm{po}}(E) \;; \qquad (2.44)$$

$$\omega^{\mathrm{osc}}(E) = \sum_{\mathrm{po}} \omega_{\mathrm{po}}(E) \quad ; \quad \omega_{\mathrm{po}}(E) = \left(\frac{\hbar}{\tau_{\mathrm{po}}(E)}\right)^{2} d_{\mathrm{po}}(E) \,. \qquad (2.45)$$

---

[7] The following discussion also holds true for sums over nonperiodic, recurrent, orbits, which will also be under consideration later.

The thermodynamic functions $D^{\mathrm{osc}}(\mu)$, $N^{\mathrm{osc}}(\mu)$, and $\Omega^{\mathrm{osc}}(\mu)$ are obtained from (2.40a)–(2.40c) by replacing the full functions by their oscillating components. The resulting integrals are of the form

$$I(T) = \int_0^\infty dE\, A(E) \exp\left[\frac{i}{\hbar}S(E)\right] f'(E - \mu) .\tag{2.46}$$

They involve the convolution of the functions $d^{\mathrm{osc}}(E)$, $n^{\mathrm{osc}}(E)$ and $\omega^{\mathrm{osc}}(E)$, oscillating around μ with a typical frequency $2\pi\hbar/\tau_{\mathrm{po}}(\mu)$, with the derivative of the Fermi factor $f'(E - \mu)$. The above integral is conveniently solved by using a Matsubara representation of the Fermi function and contour integration. To leading order in $\hbar$ and β^{-1}, but without any assumption concerning their relative values, one finds[8]

$$I(T) = I_0\, R(\tau/\tau_T) .\tag{2.47}$$

$I_0 \equiv I(T{=}0)$ is the zero-temperature result of (2.46) (with $f' = -\delta(E - \mu)$):

$$I_0 = -A(\mu) \exp\left[\frac{i}{\hbar}S(\mu)\right] .\tag{2.48}$$

The temperature dependence enters by means of the reduction factor

$$R(x) = \frac{x}{\sinh x} \quad ; \quad x = \frac{\tau}{\tau_T} .\tag{2.49}$$

Here $\tau(E)$ is the orbit period and

$$\tau_T = \frac{\beta\hbar}{\pi}\tag{2.50}$$

defines a thermal cutoff time. For systems without a potential, e.g. billiards, the period of the trajectory is related to its length L by $\tau(\mu) = L/v_F$. $R(x)$ can then be written as $R(L/L_T)$ with the thermal length L_T defined as

$$L_T = \frac{\hbar v_F \beta}{\pi} .\tag{2.51}$$

The convolution with the derivative of the Fermi function yields a damping of the oscillating periodic-orbit contributions. At very low temperature one obtains the Sommerfeld expansion

$$R(x) \simeq 1 - \frac{1}{6}x^2 .\tag{2.52}$$

For long trajectories or high temperature, R yields an exponential suppression

$$R(x) \longrightarrow 2x\, e^{-x} \qquad \text{for} \quad x \longrightarrow \infty ,\tag{2.53}$$

and therefore the only trajectories contributing significantly to thermodynamic functions are those with $\tau \lesssim \tau_T$.

---

[8] See Appendix A in [45] which is a generalization of the standard computation in [112].

By using (2.47) we get expressions for the respective thermodynamic functions in terms of the semiclassical density of states [45]:

$$D^{\mathrm{osc}}(\mu) = \sum_{\mathrm{po}} D_{\mathrm{po}}(\mu) \quad ; \quad D_{\mathrm{po}}(\mu) = R\left(\frac{\tau_{\mathrm{po}}}{\tau_T}\right) d_{\mathrm{po}}(\mu) \tag{2.54a}$$

$$N^{\mathrm{osc}}(\mu) = \sum_{\mathrm{po}} N_{\mathrm{po}}(\mu) \quad ; \quad N_{\mathrm{po}}(\mu) = R\left(\frac{\tau_{\mathrm{po}}}{\tau_T}\right)\left(-\frac{\hbar}{\tau_{\mathrm{po}}}\right) \tilde{d}_{\mathrm{po}}(\mu) \tag{2.54b}$$

$$\Omega^{\mathrm{osc}}(\mu) = \sum_{\mathrm{po}} \Omega_{\mathrm{po}}(\mu) \quad ; \quad \Omega_{\mathrm{po}}(\mu) = R\left(\frac{\tau_{\mathrm{po}}}{\tau_T}\right)\left(\frac{\hbar}{\tau_{\mathrm{po}}}\right)^2 d_{\mathrm{po}}(\mu) \,. \tag{2.54c}$$

These expressions will be used frequently throughout this work. Summarizing, they are obtained by consistently solving all intermediate integrals for the Green function, trace, etc. to leading order in $\hbar$.

Temperature smoothing suppresses the higher harmonics of the oscillating parts of thermodynamic quantities which are associated with long classical orbits in a semiclassical treatment. In contrast, finite temperature has no effects on the mean quantities (for a degenerate electron gas, $\beta\mu \gg 1$). Hence, temperature is the tuning parameter for passing from $d(E)$ at $T = 0$ to $\bar{D}(E) = \bar{d}(E)$ at large temperatures.

Equation (2.54c) provides the semiclassical approximation of the oscillating part of the grand canonical potential. In Sect. 4.2.3 we present a method to calculate the free energy and related quantities within the canonical ensemble, which turns out to be of relevance when considering mesoscopic ensemble averages.

2.4 Semiclassical Linear Response

Usually, electronic properties of mesoscopic quantum systems, such as their density of states or spectral correlations, are not directly accessible. Instead, the system's properties are typically investigated by exposing the system to external fields and studying its response as a function of external and internal parameters. In fact, a considerable number of the effects which have been observed in mesoscopic physics do not rely on strong external fields. Small test fields suffice to provide insight into a variety of mesoscopic phenomena: prominent examples include the following:

(i) Properties related to transport through systems coupled to reservoirs exhibiting a small potential difference. This includes both the Kubo bulk conductivity and conductance through phase-coherent devices in the framework of Landauer theory.
(ii) Closed quantum dots in external constant magnetic and electric fields.
(iii) The dynamic response of mesoscopic particles in frequency-dependent potentials. This includes, for example, far-infrared or high-frequency ab-

sorption in quantum dots and induced (persistent) currents in ring geometries due to a time-dependent flux.

Linear response theory has proven to be an appropriate framework for the theoretical treatment of the above-mentioned situations. The aim of this section will be to establish a connection between semiclassical approaches and linear response theory. We therefore focus on intrinsic complex effects in mesoscopic devices in a regime where linear response theory holds. The description of mesoscopic phenomena directly related to a possible nonlinear coupling due to strong external fields, for example photon-assisted transport through quantum dots [113], is therefore beyond the scope of this section.

In Chap. 3 we shall apply the general semiclassical approach [114] to dynamic response functions presented here to quantum transport in antidot superlattices.

2.4.1 Basic Quantum Mechanical Relations

There has been an ongoing discussion in the literature on the question of how to include inelastic processes in linear-response formulas for mesoscopic systems [115–117], especially if the systems are finite. For instance, the Kubo formula for the real part of the frequency-dependent conductivity reads, in the absence of any inelastic effects,

$$\mathrm{Re}[\sigma(\omega)] = \frac{2\pi e^2}{\omega V} \sum_{n,m} [f(E_n) - f(E_m)] \, |J_{nm}|^2 \, \delta(E_m - E_n - \hbar\omega). \quad (2.55)$$

V is the volume of the system and J_{nm} the matrix element of the current operator. Such a formula and its derivation are valid for an infinite system with a continuous spectrum, which allows for real transitions at energy $\hbar\omega$. For finite systems, the levels $|n\rangle$ are discrete and the response is nonzero only when $\hbar\omega$ equals exactly an energy-level difference, i.e. it is zero for almost all frequencies. It was then argued by Imry and Shiren [116] and others that, in order to achieve absorption of radiation from the field, the finite system has to be coupled to an external bath to which the absorbed energy can be transferred. This is accompanied by a finite level width γ, which then allows for real transitions to take place. As originally proposed by Czycholl and Kramer [115], it has therefore become a frequently used procedure [116,118] to introduce an inelastic width $i\gamma$ into (2.55) "by hand", leading to a broadening of the δ functions into Lorentzians

$$\delta_\gamma(\epsilon) = \frac{1}{\pi} \frac{\gamma}{\epsilon^2 + \gamma^2} \, . \quad (2.56)$$

Trivedi and Browne [117] reexamined this problem with particular emphasis on an accurate treatment of additional terms in the conductivity due to induced currents in ring structures exposed to a time-dependent field. They performed their linear-response derivation in the framework of a master equation for the reduced density matrix. This includes, besides free propagation according to the Liouville equation, a relaxation term which provides a driving of the system towards equilibrium. Such a term is essential in a linear response theory, since it provides and defines an equilibrium state as a reference from which linear deviations (of the density matrix) can be studied. The results of Trivedi and Browne exhibit the same Lorentzian behavior for the Kubo-type conductivity (2.55) as proposed above. Their approach, including terms specific to a ring topology, were recently applied by Reulet and Bouchiat [119] to the AC conductivity of disordered rings.

In all the above-mentioned approaches dissipation is included on a rather heuristic level which has proven useful and sufficient for the majority of mesoscopic phenomena. Here, we proceed along these lines and thus exclude questions concerning a more elaborate treatment of dissipation and the possible limits of linear response theory.

Before introducing our semiclassical concepts, we provide the necessary quantum mechanical dynamic response functions. We start from the dynamic susceptibility for noninteracting quasiparticles in terms of single-particle levels $|n\rangle$. Its imaginary part is given by [120]

$$R(\mu;\omega) \equiv -\mathrm{Im}\, \frac{g_\mathrm{s}}{\pi V} \sum_{n,m} \frac{[f(E_n) - f(E_m)]\, |\langle n|\hat{A}|m\rangle|^2}{E_m - E_n - \hbar\omega + i\gamma} \tag{2.57a}$$

$$= \frac{g_\mathrm{s}}{V} \sum_{n,m} [f(E_n) - f(E_m)]|A_{nm}|^2 \delta_\gamma(E_m - E_n - \hbar\omega). \tag{2.57b}$$

Here $f(E)$ is the Fermi function (2.39) and $\delta_\gamma(\epsilon)$ is a Lorentzian as defined in (2.56). The functions A_{nm} are, as usual, matrix elements $\langle n|\hat{A}|m\rangle$ of an operator $\hat{A}$ giving rise to transitions between different states. Here we focus on the imaginary part of the dynamic susceptibility since this is relevant for conductivity and absorption expressions. Introducing an additional energy integration gives

$$R(\mu;\omega) \tag{2.58}$$

$$= \frac{g_\mathrm{s}}{V} \int \mathrm{d}E \sum_{n,m} [f(E_n) - f(E_m)]|A_{nm}|^2 \delta(E - E_m)\delta_\gamma(E_m - E_n - \hbar\omega).$$

In order to take into account effects of weak disorder one has to perform averages, denoted by $\langle\ldots\rangle$, and one finds[9]

$$R_\Gamma(\mu;\omega) \equiv \langle R(\mu;\omega)\rangle \simeq g_\mathrm{s} \int \mathrm{d}E [f(E - \hbar\omega) - f(E)]\, C_\Gamma(E,\omega). \tag{2.59}$$

Here, Γ stands for a level broadening due to disorder and

$$C_\Gamma(E,\omega) \equiv \langle C_0(E,\omega)\rangle \tag{2.60}$$

---

[9] The replacement of E_n and E_m by $E - \hbar\omega$ and E in the Fermi function is possible since the inelastic broadening γ can be absorbed into the disorder broadening if the latter is, as usual, significantly larger [121].

with

$$C_0(E, \omega) = \frac{1}{V} \sum_{n,m} |A_{nm}|^2 \, \delta(E - E_m) \, \delta(E_m - E_n - \hbar\omega) \,. \qquad (2.61)$$

In the following we shall provide the equivalent response formulas in terms of traces of products of single-particle Green functions. Such representations are most appropriate for including the averaging over weak disorder. Writing the matrix elements in (2.61) in the position representation,

$$A_{nm} = \int d\boldsymbol{r} \, \psi_n^*(\boldsymbol{r}) \, A(\boldsymbol{r}) \, \psi_m(\boldsymbol{r}) \,, \qquad (2.62)$$

and using the fact that the δ functions in (2.61) can be expressed through differences between retarded and advanced Green functions (2.2), one finally has

$$C_0(E, \omega) = -\frac{1}{4\pi^2 V} [A_0^{++}(E; \omega) + A_0^{--}(E; \omega)$$

$$-A_0^{+-}(E; \omega) - A_0^{-+}(E; \omega)] \qquad (2.63a)$$

$$= -\frac{1}{2\pi^2 V} \mathrm{Re} [A_0^{++}(E; \omega) - A_0^{+-}(E; \omega)] \,. \qquad (2.63b)$$

Here the functions $A_0^{\pm\pm}$ denote traces

$$A_\Gamma^{\pm\pm}(E; \omega) \equiv \mathrm{Tr}[\hat{A} \, G^\pm(E + i\Gamma) \, \hat{A} \, G^\pm(E - \hbar\omega + i\Gamma)] \qquad (2.64a)$$

$$= \int d\boldsymbol{r}\,' \int d\boldsymbol{r} \, A(\boldsymbol{r}') \, G^\pm(\boldsymbol{r}', \boldsymbol{r}; E + i\Gamma)$$

$$\times A(\boldsymbol{r}) \, G^\pm(\boldsymbol{r}, \boldsymbol{r}'; E - \hbar\omega + i\Gamma) \qquad (2.64b)$$

for $\Gamma = 0$. The formula (2.63b) follows from (2.63a) by means of

$$G^\pm(\boldsymbol{r}, \boldsymbol{r}'; E + i\Gamma) = [G^\mp(\boldsymbol{r}', \boldsymbol{r}; E + i\Gamma)]^* \,. \qquad (2.65)$$

To account for weak disorder one has to evaluate the average $\langle \ldots \rangle$ which enters into (2.60) over products of Green functions. The semiclassical treatment of such disorder averages is described in detail in Chap. 5. Here we shall be mainly interested in spectral features in the regime of ballistic dynamics resulting from geometrical effects of the confinement potentials. We shall thus include weak residual disorder only in order to account for the related damping of signals of the clean system; we are not particularly interested in weak-localization-like effects due to disorder in the diffusive limit. Thus, vertex corrections giving rise to, for example, weak localization can be neglected and we replace $\langle G^\pm G^\pm \rangle$ by $\langle G^\pm \rangle \langle G^\pm \rangle$ in the disorder average of (2.64a).[10]

---

[10] Under the additional assumption of δ-like disorder this factorization leads, for instance, to the correct Drude conductivity, with the transport mean free path (see Appendix A.3) equal to the elastic mean free path.

The impurity average of the single-particle Green functions within the Born approximation causes a disorder broadening related to the self-energy Γ, which is assumed to be constant (see Chap. 5):

$$\langle G^{\pm}(E)\rangle \simeq G^{\pm}(E + i\Gamma) . \tag{2.66}$$

$\Gamma = \hbar/2\tau$, where τ is the single-particle relaxation time. The corresponding elastic mean free path is $l = v\tau$. With these approximations we then find for the disorder-averaged response function

$$C_{\Gamma}(E,\omega) = \langle C_0(E,\omega)\rangle = -\frac{1}{2\pi^2 V}\text{Re}[A_{\Gamma}^{++}(E;\omega) - A_{\Gamma}^{+-}(E;\omega)] , \tag{2.67}$$

with $A_{\Gamma}^{\pm\pm}$ as defined in (2.64a). Reexpressing the traces over Green functions in terms of matrix element sums by means of (2.2), one obtains

$$C_{\Gamma}(E,\omega) = \frac{1}{V}\sum_{n,m}|A_{nm}|^2 \, \delta_{\Gamma}(E - E_m) \, \delta_{\Gamma}(E_m - E_n - \hbar\omega) . \tag{2.68}$$

This relation is a generalization of (2.61) in the presence of weak disorder. It enters, via (2.59), into the dynamic response function R_{Γ}.

2.4.2 Semiclassical Approximation: Overview

The broadening Γ and the temperature average via the Fermi functions introduce a natural smoothing of the spectral properties entering into mesoscopic response functions. This is the most favorable situation for a semiclassical treatment since the related trace formulas in terms of classical phase-carrying paths then exhibit a natural cutoff. The formulas are expected to be convergent as long as the smearing is larger than the mean spacing Δ.[11]

To my knowledge, Wilkinson [122] was the first to perform a semiclassical evaluation of matrix element sums of the type given in (2.61) for the chaotic case. This type of matrix element sum has been thoroughly investigated in [86, 123–127]. Related semiclassical approaches to the static conductivity in terms of semiclassical Green functions were used in [52, 66, 67, 128]. More recently, the above semiclassical approaches have been generalized by Mehlig and Richter to a semiclassical treatment of dynamic linear response functions for integrable and chaotic ballistic quantum systems at finite frequency and finite temperature [114, 121]. This approach represents the basis of the analysis in the following section. Similar dynamic response functions were presented in [129] for the integrable case and in [130] for hyperbolic systems.

As already discussed with regard to the density of states in Sect. 2.1, an appropriate semiclassical approximation has to account for the characteristics

---

[11] On a more formal level, the exponential damping of the orbits must exceed $h/2$ (h being the topological entropy) in order to shift the regime of convergence of a semiclassical trace formula such that it is absolutely convergent for real energies.

of the classical dynamics, i.e. whether the classical motion of a mesoscopic quantum system is regular or chaotic. Nevertheless, semiclassical dynamic response formulas exhibit a similar overall structure in both cases.

First, to lowest order in $\hbar$ they contain a purely classical part (corresponding to the Weyl part of the density of states) varying smoothly as a function of energy or other external parameters. This will be expressed in terms of averages over correlation functions of the Weyl symbols, the classical analogs of the operators $\hat{A}$. For chaotic systems this average is just the microcanoncial average over the energy shell; for integrable systems the average ranges over invariant tori.

Secondly, quantum fluctuations are expressed, to leading order in $\hbar$, through trace formulas over periodic-orbit contributions. The periodic orbits consist of isolated unstable orbits for chaotic systems and representative periodic paths on resonant tori for integrable systems, respectively. Their $\hbar$ dependence is different for chaotic and integrable systems.

One can follow two equivalent approaches for a semiclassical evaluation of response functions of the form of $R_\Gamma(\mu; \omega)$: either by starting from (2.68), which then involves a semiclassical computation of matrix element sums [122, 126], or by using the representation (2.67) in terms of Green functions and evaluating the trace integrals (2.64a) semiclassically. We pursue the latter approach for systems with chaotic dynamics and then briefly illustrate the former for the integrable case at the end of this section.

2.4.3 Chaotic Case

Smooth Part

A semiclassical representation of the disorder-averaged response function C_Γ, (2.67), is obtained by introducing into the trace integrals $A_\Gamma^{+-}, A_\Gamma^{++}$ in (2.64b) the semiclassical expressions (2.3) for the Green functions $G^+(r', r)$ and $G^-(r, r')$ or $G^{+*}(r', r)$, respectively (using (2.65)). The Green function products gives rise to double sums over trajectories t and t' between r and r'.

For diagonal pairs of identical paths $t = t'$ entering into A_Γ^{+-} the phases cancel up to a remaining phase $\omega \tau_t$. These pairs constitute the *classical* contribution C_Γ^0 to the response function. Combining the prefactors, we have

$$C_\Gamma^0(E, \omega) = \frac{1}{4(\pi\hbar)^3 V} \mathrm{Re} \int dr\, dr'\, A(r)\, A(r')$$

$$\times \sum_{t=t'} \begin{vmatrix} \dfrac{\partial^2 S_t}{\partial r \partial r'} & \dfrac{\partial^2 S_t}{\partial r \partial E} \\[2mm] \dfrac{\partial^2 S_t}{\partial E \partial r'} & \dfrac{\partial^2 S_t}{\partial E^2} \end{vmatrix} \exp\left(i\omega\tau_t - \frac{\tau_t}{\tau}\right). \tag{2.69}$$

At this point it is convenient [66, 67, 128] to transform the r' integral into an integral over initial momenta as in (2.6): for a given initial point r one may either perform the spatial integration over r' or integrate over all initial momenta p on the energy shell $\delta[E - \mathcal{H}(r,p)]$ and, for a given direction, along the then uniquely defined trajectory. The latter integration can be represented as a time integral which additionally involves the Weyl symbols $A[r(t)]$, $A[r'(t')]$. The Jacobian of this transformation of the integration variables is exactly provided by the determinant in (2.69), the square of the classical phase space density. As the result one finds

$$C_\Gamma^0(E,\omega) = \frac{1}{4(\pi\hbar)^3 V}$$

$$\times \int d\mathbf{r}\, d\mathbf{p} \int_0^\infty dt\; A(0)A(t)\; \cos(\omega t)\; e^{-t/\tau} \delta[E - \mathcal{H}(\mathbf{r},\mathbf{p})]\;. \qquad (2.70)$$

This is conveniently expressed as

$$C_\Gamma^0(E,\omega) = \frac{\bar{d}(E)}{(\pi\hbar)V g_\mathrm{s}} \int_0^\infty dt \langle A(0)A(t)\rangle_{\mathbf{pr}}\; \cos(\omega t)\; e^{-t/\tau}\;, \qquad (2.71)$$

where we have introduced the microcanonical phase space average (for chaotic systems)

$$\langle B\rangle_{\mathbf{pr}} \equiv \frac{g_\mathrm{s}}{\hbar^2 \bar{d}(E)} \int d\mathbf{r}\, d\mathbf{p}\; B(\mathbf{r},\mathbf{p})\; \delta[E - H(\mathbf{r},\mathbf{p})]\;. \qquad (2.72)$$

In the above equations $\bar{d}(E)$ is the mean density of states including spin. C_Γ^0 is usually a smooth function of energy or other external parameters. It represents the leading-order $\hbar$ contribution to the smooth part of C_Γ, similarly to the Weyl part, the leading-order term for the smooth part of the density of states.[12] In the semiclassical limit the classical term C_Γ^0 yields the dominant contribution to the overall response function C_Γ.

The smooth part of the dynamic response function, $R_\Gamma^0(\mu;\omega)$, is obtained by integrating over the Fermi functions. Assuming $C_\Gamma^0(E;\omega) \simeq C_\Gamma^0(E+\hbar\omega;\omega)$, we can expand the Fermi function in the integral (2.59) according to $f(E - \hbar\omega) \simeq f(E) + \hbar\omega\delta(E - \mu)$ and find [114]

$$R_\Gamma^0(\mu,\omega) \simeq \omega \frac{\bar{d}(\mu)}{\pi V} \int_0^\infty dt \langle A(0)A(t)\rangle_{\mathbf{pr}}\; \cos(\omega t)\; e^{-t/\tau}\;. \qquad (2.73)$$

Quantum Oscillations

In order to account for quantum mechanical contributions to response functions one has to include nondiagonal terms corresponding to different paths. For contributions of pairs of paths to $A_\Gamma^{++}(E;\omega)$ and of pairs of different paths to $A_\Gamma^{+-}(E;\omega)$, a net phase remains in C_Γ, (2.67): these terms give rise to quantum corrections to the Weyl part owing to interference effects.

---

[12] We note, however, that the Weyl term can be related to "paths of zero length", while C_Γ^0 stems from finite-length trajectories.

Both trace integrals can be computed in the stationary-phase approximation assuming chaotic classical dynamics. The details of the calculation for
$A_\Gamma^{++}(E;\omega)$ and $A_\Gamma^{+-}(E;\omega)$ can be found in Appendix A.1.

(i) Contributions from G^+G^- Terms. The oscillatory part of $A_\Gamma^{+-}(E;\omega)$
depends on products of advanced and retarded semiclassical Green functions
related to different path lengths. The resulting expression reads (see Appendix A.1.3, (A.30))

$$(A_\Gamma^{+-})^{\mathrm{osc}}(E;\omega) \simeq \frac{1}{\hbar^2} \sum_{\mathrm{po}} \mathcal{C}_{\mathrm{po}}^{+-} \tau_{\mathrm{po}} \tag{2.74}$$

$$\times \sum_{j=1}^\infty B_{\mathrm{po}}^{(j)} \exp\left[j \frac{\tau_{\mathrm{po}}}{2} \left(i\omega - \frac{1}{\tau} \right) \right] \cos\left[j \left(\frac{S_{\mathrm{po}}}{\hbar} - \frac{\omega\tau_{\mathrm{po}}}{2} - \frac{\eta_{\mathrm{po}}\pi}{2} \right) \right].$$

The sum runs over all unstable primitive periodic orbits of the system.
$S_{\mathrm{po}}(E) = \oint_{\mathcal{C}_{\mathrm{po}}} p\, dq$ is the classical action, τ_{po} the period, and η_{po} the Morse
index of each periodic orbit. The index j counts higher repetitions. The pref-
actor

$$B_{\mathrm{po}}^{(j)} = \left| \det(M_{\mathrm{po}}^j - 1) \right|^{-1/2} \tag{2.75}$$

accounts for the classical (in)stability of the orbits given by the monodromy
matrix M_{po}. The functions $\mathcal{C}_{\mathrm{po}}^{+-}$ denote Fourier transforms

$$\mathcal{C}_{\mathrm{po}}^{+-}(E;\omega) \equiv \int_0^\infty dt\, e^{i\omega t - t/\tau}\, \tilde{C}_{\mathrm{po}}(t) \tag{2.76}$$

of classical autocorrelation functions

$$\tilde{C}_{\mathrm{po}}(t) = \frac{1}{\tau_{\mathrm{po}}} \int_0^{\tau_{\mathrm{po}}} dt'\, A(t + t') A(t') \tag{2.77}$$

of $A(r)$ along each prime periodic orbit. The weak-disorder average leads to
an exponential damping[13] of each periodic-orbit contribution on the scale of
$j\tau_{\mathrm{po}}/2\tau$.

Except for the correlation functions, the formula (2.74) has the same
structure as Gutzwiller's trace formula (2.27) for the density of states. However, here the periodic-orbit contributions represent pairs of nonclosed paths
(up to infinite length) both lying on the same periodic orbit: the trajectories
begin and end at the same point on the periodic orbit but differ in j, by their
number of full traversals around the periodic orbit. The quantum oscillations
appearing in $(A_\Gamma^{+-})^{\mathrm{osc}}(E;\omega)$ are thus related to interference between open
paths with phase differences $\sim j S_{\mathrm{po}}/\hbar$.

---

[13] A refined semiclassical treatment of the disorder average in Chap. 5 shows in fact
that the number of repetitions enters quadratically ($\sim j^2$) into the exponent. This
is of minor importance in the present calculation, since temperature provides the
predominant cutoff.

(ii) **Contributions from G^+G^+ Terms.** Quantum oscillations arising from the trace over the product of two retarded (or advanced) Green functions are given, to leading order in $\hbar$, by (see Appendix A.1.2, (A.21))

$$(A_\Gamma^{++})^{\mathrm{osc}}(E;\omega) \tag{2.78}$$

$$\simeq -\frac{1}{\hbar^2}\sum_{\mathrm{po}}\tau_{\mathrm{po}}\sum_{j=1}^{\infty}\mathcal{C}_{\mathrm{po}}^{++}\,B_{\mathrm{po}}^{(j)}\exp\left(-j\frac{\tau_{\mathrm{po}}}{2\tau}\right)\exp\left[ij\left(\frac{S_{\mathrm{po}}}{\hbar}-\frac{\eta_{\mathrm{po}}\pi}{2}\right)\right].$$

In this case the Fourier-like transform of the orbit correlation function is defined as

$$\mathcal{C}_{\mathrm{po}}^{++}(E;\omega) = \int_0^{j\tau_{\mathrm{po}}}dt\,e^{i\omega t}\tilde{C}_{\mathrm{po}}(t)\,. \tag{2.79}$$

In the static limit this takes the form

$$\mathcal{C}_{\mathrm{po}}^{++}(E;\omega\equiv 0) = \frac{1}{\tau_{\mathrm{po}}}\langle A\rangle_{\mathrm{po}}^2\,, \tag{2.80}$$

where $\langle A\rangle_{\mathrm{po}}$ is the average of $A(\boldsymbol{r})$ along the periodic orbit. In many applications this average is indeed zero, see e.g. Chap. 3, and the G^+G^+ contribution disappears in the static case. However, it should be pointed out that, contrary to diffusive dynamics, physically relevant contributions from G^+G^+ terms may exist in the ballistic regime, especially at finite ω.

(iii) **Trace Formula.** To elucidate the role of periodic orbits in the response at *finite* frequencies it proves convenient to expand the autocorrelation function $\tilde{C}_{\mathrm{po}}$ in (2.77), which is periodic in τ_{po}, into a Fourier series

$$\tilde{C}_{\mathrm{po}}(t) \equiv \sum_{m=-\infty}^{\infty}\alpha_{\mathrm{po}}^{(m)}\,e^{-i\omega_m t}\qquad\text{with}\qquad \omega_m = \frac{2\pi m}{\tau_{\mathrm{po}}}\,. \tag{2.81}$$

The Fourier coefficients $\alpha_{\mathrm{po}}^{(m)}$ are real and $\alpha_{\mathrm{po}}^{(0)}=0$ owing to the symmetry of the correlation functions. One then finds, from (2.76) and (2.79) [121],

$$\mathcal{C}_{\mathrm{po}}^{+-}(E,\omega) = i\sum_{m=-\infty}^{\infty}\frac{\alpha_{\mathrm{po}}^{(m)}}{\hbar(\omega-\omega_m)+2i\Gamma}\,, \tag{2.82a}$$

$$\mathcal{C}_{\mathrm{po}}^{++}(E,\omega) = 2\sin\left(j\frac{\omega\tau_{\mathrm{po}}}{2}\right)\exp\left(-ij\frac{\omega\tau_{\mathrm{po}}}{2}\right)\sum_{m=-\infty}^{\infty}\frac{\alpha_{\mathrm{po}}^{(m)}}{\hbar(\omega-\omega_m)}\,. \tag{2.82b}$$

Upon combining the real parts of $(A_\Gamma^{+-})^{\mathrm{osc}}(E;\omega)$, (2.74), and $(A_\Gamma^{++})^{\mathrm{osc}}(E;\omega)$, (2.78), one obtains for the quantum corrections to the disorder-averaged response function $C_\Gamma(E,\omega)$ ((2.67)), to leading order in $\hbar$,

$$C_\Gamma^{\mathrm{osc}}(E,\omega) \tag{2.83}$$

$$\simeq \frac{-1}{\pi^2\hbar^2 V}\sum_{\mathrm{po}}\tau_{\mathrm{po}}\sum_{j=1}^{\infty}B_{\mathrm{po}}^{(j)}\exp\left(-j\frac{\tau_{\mathrm{po}}}{2\tau}\right)\cos\left[j\left(\frac{S_{\mathrm{po}}}{\hbar}-\frac{\pi\eta_{\mathrm{po}}}{2}-\frac{\omega\tau_{\mathrm{po}}}{2}\right)\right]$$

$$\times\sum_{m=-\infty}^{\infty}\alpha_{\mathrm{po}}^{(m)}\frac{2\Gamma}{[\hbar(\omega-\omega_m)]^2+(2\Gamma)^2}\left[\cos\left(\frac{j\omega\tau_{\mathrm{po}}}{2}\right)-\frac{\sin\left(j\omega\tau_{\mathrm{po}}/2\right)}{(\omega_m-\omega)\tau}\right].$$

For the calculation of the corresponding oscillatory part of the response function $R_\Gamma(\mu; \omega)$, (2.59), the related integral over the Fermi function has to be evaluated. The classical contribution is obtained by expanding the Fermi function as $f(E - \hbar\omega) \simeq f(E) + \hbar\omega\delta(E - \mu)$. We point out that this expansion yields incorrect results when simply applied to the oscillatory contributions.[14] Instead, the integral can be evaluated analytically in a leading-order $\hbar$ approximation using the relation (2.47). Together with the classical "Weyl" contribution (2.73) the overall semiclassical dynamic response is then given by [114]

$$R_\Gamma(\mu; \omega) \simeq R_\Gamma^0(\mu; \omega) + R_\Gamma^{\mathrm{osc}}(\mu; \omega)$$

$$\simeq R_\Gamma^0(\mu, \omega) - \frac{2g_s}{\pi\hbar V} \sum_{\mathrm{po}} \sum_{j=1}^{\infty} B_{\mathrm{po}}^{(j)} \, R\left(\frac{j\tau_{\mathrm{po}}}{\tau_T}\right) \exp\left(-j\frac{\tau_{\mathrm{po}}}{2\tau}\right)$$

$$\times \cos\left[j\left(\frac{S_{\mathrm{po}}(\mu)}{\hbar} - \frac{\pi\eta_{\mathrm{po}}}{2}\right)\right] \sin\left(\frac{j\omega\tau_{\mathrm{po}}}{2}\right)$$

$$\times \sum_{m=-\infty}^{\infty} \left\{\alpha_{\mathrm{po}}^{(m)} \frac{2\Gamma}{[\hbar(\omega - \omega_m)]^2 + [2\Gamma]^2}\right.$$

$$\left. \times \left[\cos\left(\frac{j\omega\tau_{\mathrm{po}}}{2}\right) - \frac{\sin\left(j\omega\tau_{\mathrm{po}}/2\right)}{(\omega_m - \omega)\tau}\right]\right\}. \tag{2.84}$$

This represents the main result of this section. In a semiclassical representation, the dynamic susceptibilities are composed of a (dominant) classical part and oscillatory quantum corrections. Equation (2.84) establishes for both the smooth part R_Γ^0 and the oscillatory part R_Γ^{osc} the leading-order semiclassical expansion of a dynamic response function of the type given in (2.59). The expression provides a means to calculate or analyze transport coefficients or susceptibilities in the semiclassical regime. It accounts for finite frequency and finite temperature in the presence of weak disorder.

We complete this section with a few further remarks concerning the properties and limitations of the semiclassical response formula.

The Lorentzian functions entering into the oscillatory part make the underlying physical processes rather transparent: if the frequency ω of the external time-dependent field is close to the eigenfrequency $\omega_m = 2\pi m/\tau_{\mathrm{po}}$ of a periodic orbit, the response function exhibits a resonance, indicating, for example, a peak in the absorbed power.

The impurity average gives rise to an exponential damping on a time scale of τ. For large Γ, corresponding to $\tau \ll \tau_{\mathrm{po}}$, the periodic-orbit contributions are damped out. On the other hand, the semiclassical approximations are only reliable as long as $\Delta < \Gamma$.

We note that in the limit of small Γ one can easily verify that the oscillatory contributions vanish to lowest order in $\hbar$ [121]. Hence in *clean* systems one does *not* expect periodic-orbit contributions to the dynamic response at

---

[14] This procedure has apparently been used in [130].

this level of approximation. This fact stems from the difference of the Fermi functions and is not observed in the literature that deals with pure matrix element formulas.

In (2.84), temperature effects are included through the function $R(x) = x/\sinh x$ and give rise to an exponential damping of contributions from orbits with $\tau_{\text{po}} > \tau_T = \hbar\beta/\pi$, see (2.50). Hence, typically only a few periodic orbits suffice to describe the quantum response in the mesoscopic regime. At temperatures such that τ_T is smaller than the period of the shortest periodic orbit, the quantum oscillations disappear and only the classical contribution remains.

The semiclassical derivation of (2.84) relies on the expansion $S_{\text{po}}(E + \hbar\omega) \simeq S_{\text{po}}(E) + \hbar\omega\tau_{\text{po}}$. Hence, this semiclassical expression is applicable for frequencies such that $\hbar\omega \ll \mu$. This is the usual situation in mesoscopic systems in the semiclassical regime. The expression is generally not suited to describe, for example, frequency-dependent excitations from low-lying states in atoms or molecules. For such applications, corresponding semiclassical methods have been developed which appropriately take into account the coupling between a well-localized ground state and an extended excited state [131].

2.4.4 Integrable Case

The integrable case differs from the chaotic one with regard to the methods used to evaluate the trace integrals in the response formula (2.67): the stationary-phase approximation as was used in Appendix A.1 for the case of isolated orbits is not applicable. However, the overall final response formula is of similar structure and we briefly summarize the result. It can be obtained by making contact with existing semiclassical approximations for matrix element expressions for integrable classical dynamics without disorder, which are of the form (2.61) for C_0.

To this end one first expresses C_Γ, which enters into the integral (2.59) of the response function R_Γ, in terms of C_0 by means of two convolutions [114]:

$$C_\Gamma(E,\omega) = \frac{1}{V} \sum_{n,m} |A_{nm}|^2 \, \delta_\Gamma(E - E_m) \, \delta_\Gamma(E_m - E_n - \hbar\omega) \tag{2.85}$$

$$= \int d\epsilon \int d\omega' \, \delta_\Gamma(E - \epsilon) \, \delta_\Gamma[\epsilon - E + \hbar(\omega - \omega')] \, C_0(\epsilon; \omega') \, .$$

Semiclassical expansions for $C_0(E;\omega)$ have been studied for the chaotic case in [86, 123, 124] and for the integrable case in [126]. Here, we cite the leading-order semiclassical approximation of the response function C_0 for the latter case:

$$C_0(E;\omega) \simeq C_0^0(E;\omega) \tag{2.86}$$

$$+ \frac{1}{\pi\hbar^{3/2}} \text{Re} \left\{ \sum_M C_M \, \tau_M \, B_M \exp\left[ij \left(\frac{S_M}{\hbar} - \frac{\eta_M \pi}{2} + \gamma \frac{\pi}{4} \right) \right] \right\} \, .$$

Note the close correspondence to the Berry–Tabor trace formula (2.25) for the density of states. The sum in (2.86) is over rational tori, labeled by the topology vector $\boldsymbol{M}$. The $B_{\boldsymbol{M}}$ are the same as in the Berry–Tabor formula. The functions $C_0^0(E; \omega)$ and $C_{\boldsymbol{M}}(E; \omega)$ are in the present case Fourier transforms of classical autocorrelation functions on tori [126], similar to those discussed for the chaotic case.

The dynamic response function R_Γ is obtained by evaluating the integrals on the right-hand side of (2.85) and the Fermi integral (2.59) to leading order in $\hbar$ by complex contour integration. As a result, one recovers the structure of the semiclassical response function (2.83) for the chaotic case [114, 121], although the classical entries are now those for the integrable case.

After these rather technical sections we now turn to applications of this semiclassical approach to dynamic susceptibilities, and then to transport in the next chapter.

2.4.5 Dynamic Susceptibilities

The absorption of radiation and the polarizability of small conducting particles represent one field of direct application[15] of the semiclassical linear-response formalism. These systems played an important role in the early, seminal work by Gorkov and Eliashberg [135] on quantum effects in the polarizability. Their work can be regarded as pioneering in the context of quantum chaos: it already invokes random-matrix theory to account for the effect of different spectral statistics on the polarizability. Absorption of radiation provides a further means to study quantum chaos in mesoscopic physics. Contrary to transport, one probes finite closed systems and their optical properties. Today's optical experiments on mesoscopic systems span a wide range from atomic metal clusters to high-mobility semiconductor devices of reduced dimensionality like quantum dots. They allow the observation of the quantum size effects mentioned above (e.g. [136–138]).

In [114, 139] the semiclassical dynamic response function $R_\Gamma(\mu; \omega)$, (2.84), has been employed to compute the far-infrared absorption in small ballistic particles. Here we summarize only the main results. The coefficient $\alpha(\omega)$ for frequency-dependent absorption of radiation is directly related to $R_\Gamma(\mu; \omega)$ via $\alpha(\mu; \omega) = (\pi e^2 \omega V / 2E_0^2) R_\Gamma(\mu; \omega)$, where E_0 denotes the amplitude of the external field. Hence the absorption coefficient can be decomposed into

$$\alpha(\mu; \omega) \simeq \alpha^0(\mu; \omega) + \alpha^{\mathrm{osc}}(\mu; \omega) . \tag{2.87}$$

In the approach of [114] the effect of Coulomb interactions has been consistently incorporated into the semiclassical framework by using a Thomas–Fermi approximation for frequencies below the plasma frequency [140]. Then

---

[15] Reviews of the extensive literature documenting the long history of studies on the polarizability and absorption of small particles can be found, for example, in [132–134].

the screened effective potential of the radiation field enters as a perturbation $A(r)$ into the response functions R_Γ. In [114] $\alpha(\mu; \omega)$ has been calculated for the case of two-dimensional disks both semiclassically and quantum mechanically. Semiclassics predicts a classical optical absorption profile $\alpha^0(\mu; \omega)$ with pronounced peaks at frequencies where the classical correlation functions exhibit resonances. On these are superimposed lower-order (in $\hbar$) oscillatory quantum corrections α^{osc}. The semiclassical results are in very good quantitative agreement with the numerical quantum results for different temperatures and (weak) disorder damping. Details can be found in [114].

3. Ballistic Quantum Transport

The study of electronic transport through small conductors is one of the most prominent research areas in mesoscopic physics. The particular interest in transport during most of the last two decades can be related to the fact that mesoscopic devices show a variety of effects which may appear surprising from a macroscopic point of view. These phenomena are related to quantum interference at low temperatures and therefore give rise to nonohmic behavior. Phase coherence effects were first observed in transport through metal conductors: a well-known example is conductance fluctuations in disordered samples. Since the variance of these oscillations is practically independent of system size, disorder, or material, they are termed *universal* [2]. A second outstanding phenomenon is a quantum correction to the averaged conductance, the *weak-localization* effect, and its magnetic-field dependence.

Mesoscopic transport was orginally focused on small metal devices with diffusive electron motion. However, with the advent and development of high-mobility semiconductor heterostructures the regime of ballistic transport became accessible. In this limit impurity scattering is strongly reduced. Hence, the electron dynamics can be controlled by lithographically imposing additional electrostatic potential barriers. If the elastic mean free path is considerably larger than the system size the conductance will reflect the geometry of the microstructure. The question which naturally arises is one of quantum phenomena in the ballistic regime and their relation to the interference of electron waves multiply reflected in the artificial potential landscapes of such systems [3].

Progress in ballistic microcavities, moreover oriented interest towards questions of how electron dynamics that is classically regular or chaotic shows up in the properties of ballistic quantum transport [42,51,55,56]. Three types of experiments have been generally used to investigate this issue:

(i) A whole sequence of experiments deals with phase-coherent transport through microconductors, lithographically designed as billiard-like electron cavities, attached to leads [44,55,68,141–150]. These "quantum billiards", an example of which was given in the introduction, enable one to study the effect of the geometry on quantum transport. It was shown, both theoretically and experimentally, that chaotic ballistic cavities exhibit a universal behavior for the conductance fluctuations and weak localization, characterized by a single scale. On the contrary integrable structures show nongeneric behavior.

(ii) In a second class of experiments, electron transport through artificial laterally defined superlattices studied [51, 151–156] (see the example in the introduction). These *antidot lattices*, which consist of arrays of periodically arranged repulsive pillars, can be regarded as macroscopic since they are not entirely phase-coherent. Their size is much larger than the phase coherence length ℓ_ϕ. However, the lattice constant, the relevant internal length scale, is typically of order of 50–300 nm and therefore much smaller than both the elastic mean free path l and ℓ_ϕ. This enables the observation of quantum effects in these systems also as will be discussed in this section. The combined potential of the superlattice and an external magnetic field H gives rise to a variety of peculiar phenomena, including effects of the classically chaotic electron motion.

(iii) More recently, a third class of transport experiments, which address questions related to chaos in mesoscopic quantum systems, has been performed. In these experiments, transport through relatively large quantum wells, weakly coupled through tunneling barriers to the leads, has been studied [157]. An additional, tilted magnetic field gives rise to a mixed classical phase space structure inside these resonant tunneling diodes. The measurements show oscillations in the tunneling current which clearly reflect periodic-orbit effects. In particular, states localized along short periodic orbits which connect the two tunnel barriers play a prominent role in the electron transport. The original semiclassical approach (for a review see [158]) has been refined in the meantime by a number of different groups. In [159, 160] the current has been calculated using a semiclassical evaluation of Bardeen-type tunneling matrix elements [161]. The results again point towards the importance of certain types of periodic orbits. Furthermore the role of bifurcations [162] and complex "ghost" periodic orbits [163] has been pointed out in order to achieve a quantitative understanding of the experimental results. For more details on this specific branch of semiclassical transport we refer the reader to the publications cited above and references therein.

Interference phenomena in quantum electron transport through microstructures are usually described theoretically within two complementary frameworks. The Landauer–Büttiker formalism [3] describes transmission through single phase-coherent devices. The current is directly related to transmission properties of the sample and expressed in terms of conductance coefficients between channels in the various attached leads. On the other hand, the Kubo linear response theory [120] has proved useful in the treatment of bulk transport properties of samples with a size exceeding the phase-breaking length.

A semiclassical picture relating wave interference to possibly complicated boundary-reflected paths in the ballistic regime seems physically appealing. In this spirit a semiclassical approach to the conductance within the Landauer–Büttiker framework has been used in [42, 43, 56], expressing the conductance coefficients in terms of interfering electron paths. This approach will be discussed in Sect. 3.2.1.

The more recent experiments on magnetotransport in antidot structures revealed the lack of a corresponding semiclassical approximation for the Kubo bulk conductivity [51, 144]. As a result, such an approach has been independently developed by Richter [66] and by Hackenbroich and von Oppen [67]. It can be regarded as a specific (static) case of the semiclassical linear response formalism presented in the previous section and will be described below. In particular, we focus on its use to describe quantum transport in superlattices, which can be considered as the main application so far.

We note that a quasiclassical approach to transport in disordered systems was originally proposed by Chakravarty and Schmid [7]. They studied interference between diffusive electron paths in random δ-like potentials, focusing especially on the description of weak-localization mechanisms. More recently, a comprehensive semiclassical treatment of one-dimensional disordered systems, including aspects of localization, has been presented by Dittrich [77].

3.1 Bulk Conductivity

We first present the semiclassical approach to quantum transport. After a brief account of the main experimental findings for antidot lattices we then use the semiclassical formalism to describe these experiments and summarize the corresponding quantum calculations.

3.1.1 Semiclassical Approach

As the first application of the semiclassical linear response formulas presented in Sect. 2.4 we derive semiclassical expressions for the magnetoconductivity tensor within the Kubo formalism [66, 67]. A related semiclassical evaluation of the Kubo conductivity with emphasis on weak-localization effects has been performed by Argaman [128].

Within the Kubo formalism the static conductivity for a Fermi energy μ and temperature T is given by

$$\sigma_{ij}(\mu, H; T) = g_{\rm s} \int \left(-\frac{{\rm d}f}{{\rm d}E} \right) \langle \sigma_{ij}(E, H) \rangle \, {\rm d}E \ . \tag{3.1}$$

$f(E)$ is the Fermi function (2.39) and $\langle \dots \rangle$ denotes an average over weak residual disorder. The diagonal and Hall conductivities are conveniently expressed as [167]

$$\sigma_{xx} = \frac{\pi e^2 \hbar}{V} \text{Tr} \left\{ \hat{v}_x \; \hat{\delta}(E - \hat{H}) \; \hat{v}_x \; \hat{\delta}(E - \hat{H}) \right\} , \tag{3.2a}$$

$$\sigma_{xy} = \frac{e}{V} \frac{\partial n(E, H)}{\partial H} \tag{3.2b}$$

$$+ \frac{i}{2} \frac{e^2 \hbar}{V} \text{Tr} \left\{ \hat{v}_x G^+(E) \hat{v}_y \; \hat{\delta}(E - \hat{H}) - \hat{v}_x \; \hat{\delta}(E - \hat{H}) \hat{v}_y \; G^-(E) \right\} .$$

Here, the $\hat{v}_i$ are velocity operators, and $n(E)$ the number of states below E (see (2.36)).

Chaotic Systems

We begin with the evaluation of the diagonal conductivity. Neglecting vertex corrections, we replace the disorder-average of σ_{xx} in (3.1) by the product of disorder averaged $\hat{\delta}$ operators in (3.2a) (see also the discussion preceding (2.67)). Then we have

$$\sigma_{xx}(\mu, H; T) = \frac{g_s \pi e^2 \hbar}{V} \int dE \left(-\frac{df}{dE} \right) \text{Tr} \left\{ \hat{v}_x \; \hat{\delta}_\Gamma(E - \hat{H}) \hat{v}_x \; \hat{\delta}_\Gamma(E - \hat{H}) \right\} . \tag{3.3}$$

This equation is, up to prefactors, equivalent to the static limit of the response function R_Γ given by (2.59) and (2.68). Accordingly, the semiclassical expression for the diagonal conductivity is conveniently decomposed into

$$\sigma_{xx}(\mu, H) = \sigma^0_{xx}(\mu, H) + \sigma^{\text{osc}}_{xx}(\mu, H) . \tag{3.4}$$

The smooth classical part, yielding the leading-order $\hbar$ contribution, reads, according to (2.71),

$$\sigma^0_{xx}(\mu, H) = \frac{e^2 \bar{d}(\mu)}{V} \int_0^\infty dt \; \langle v_x(t) v_x(0) \rangle_{\boldsymbol{pr}} \exp(-t/\tau) . \tag{3.5}$$

The microcanonical phase space average $\langle \ldots \rangle_{\boldsymbol{pr}}$ is defined in (2.73). τ is the elastic scattering time[1] and $\bar{d}(\mu)$ is the mean density of states at the Fermi energy; $\bar{d}(\mu) = g_s V/(2\pi\hbar^2)$ in two dimensions. The expression (3.5) represents the generalization of the Drude conductivity to systems with arbitrary chaotic phase space dynamics.

To compute the leading-order quantum corrections σ^{osc}_{xx} to the *static* conductivity, it is more convenient to employ the semiclassical expressions (2.74) and (2.78), which enter into C_Γ and hence R_Γ, than to use the final equation (2.84) for the dynamic response. For the case of the static conductivity, the contribution A^{++}_Γ vanishes since the correlation function (2.80) turns out to be zero: $\langle v_x \rangle_{\text{po}} \equiv 0$. Therefore, the quantum correction to σ_{xx} is given solely by $(A^{+-}_\Gamma)^{\text{osc}}$. Combining the prefactors, it reads [66, 67]

---

[1] A proper inclusion of ladder diagrams in the treatment of disorder would give a damping according to $\tau_{\text{tr}} = v_F/l_T$, where l_T is the transport MFP (see Appendix A.3).

$$\sigma_{xx}^{osc}(\mu, H) = \frac{2g_s}{V}\frac{e^2}{h} \tag{3.6}$$

$$\times \sum_{po} \tau_{po} \mathcal{C}_{xx}^{po} \sum_{j=1}^{\infty} \frac{R(j\tau_{po}/\tau_T)\exp(-j\tau_{po}/2\tau)}{\left|\det(M_{po}^{j}-1)\right|^{1/2}} \cos\left[j\left(\frac{S_{po}}{\hbar} - \eta_{po}\frac{\pi}{2}\right)\right].$$

Quantum corrections to the Kubo conductivity are semiclassically expressed in terms of a sum over contributions from classical periodic orbits (po) and their higher repetitions j. Temperature enters via the function $R(x) = x/\sinh(x)$, (2.49), with $\tau_T = \hbar\beta/\pi$, (2.50). The weights

$$\mathcal{C}_{xx}^{po} = \frac{1}{\tau_{po}} \int_0^\infty dt\, e^{-t/\tau} \int_0^{\tau_{po}} dt'\, v_x(t')\, v_x(t+t') \tag{3.7}$$

are velocity correlation functions along the periodic orbits. Apart from the $\mathcal{C}_{xx}^{po}$ the formula (3.6) for σ_{xx}^{osc} is essentially the same as Gutzwiller's trace formula (2.27) for the density of states.

The Hall conductivity (3.2b) in the representation by Středa [167] is decomposed into a magnetization term and a term similar to σ_{xx}. The former is obtained in the leading-order semiclassical approximation by using (2.54b) for the thermodynamic particle number and applying the H-field derivative only to the rapidly varying phase. The semiclassical evaluation of the second contribution to σ_{xy} follows by proceeding precisely along the same lines as for σ_{xx}^{osc}. The smooth contribution to σ_{xy} is of the same structure as (3.5) but for velocity correlations between v_x and v_y. The oscillatory part of the semiclassical Hall conductivity reads, to lowest order in $\hbar$ [66],

$$\sigma_{xy}^{osc}(\mu, H) = \frac{2g_s}{V}\frac{e^2}{h} \sum_{po} \sum_{j=1}^{\infty} \left(\frac{1}{e}\frac{\partial S_{po}}{\partial H} + \tau_{po}\mathcal{C}_{xy}^{po}\right)$$

$$\times \frac{R\left(j\tau_{po}/\tau_T\right)e^{-j\tau_{po}/2\tau}}{\left|\det(M_{po}^{j}-1)\right|^{1/2}} \cos\left[j\left(\frac{S_{po}}{\hbar} - \eta_{po}\frac{\pi}{2}\right)\right]. \tag{3.8}$$

Frequency-Dependent Conductivity

The semiclassical approach to the static magnetoconductivity can be generalized to finite frequencies on the basis of the semiclassical dynamic response formalism presented in Sect. 2.4. Using the relation $\sigma(\omega) = \pi e^2 R_\Gamma(\omega)/\omega$, the frequency-dependent conductivity is given, according to (2.84), to leading order in the smooth and oscillatory contributions as

$$\sigma_{ij}(\mu;\omega) = \sigma_{ij}^0(\mu;\omega) + \sigma_{ij}^{osc}(\mu;\omega). \tag{3.9}$$

This relation represents a generalization of related formulas for cyclotron resonance in the bulk. Resonance phenomena will typically arise if the frequency ω is close to one of the frequencies of the (quasi)periodic motion involved. Recently, a *classical* frequency-dependent Kubo formula based on

σ_{xx}^0 has been employed to describe microwave photoconductivity measured in antidot arrays [168]. In such experiments the magnetoconductivity signal is studied as a function of the frequency of the microwave radiation applied. The observed deviations from the two-dimensional magnetoplasmon resonance curve in the presence of an antidot lattice can be related to the underlying classical mechanics [168]. To my knowledge, resonance phenomena in *quantum* corrections $\sigma_{ij}^{osc}(\mu;\omega)$ have not yet been experimentally detected.

Shubnikov–de Haas Oscillations

Before we apply the above results to describe the magnetoconductivity anomalies of antidot lattices, we note the related expression for the conductivity of a free 2DEG. In the unmodulated case, in the presence of a homogeneous perpendicular magnetic field, the electrons perform cyclotron motion with frequency $\omega_{cyc} = eH/m^*c$. Their dynamics is integrable. Hence, the above trace formula in terms of isolated periodic orbits does not directly apply. However, we can make use of the Berry–Tabor-like linear-response expressions (2.86) derived for the integrable case in Sect. 2.4.4. The corresponding computation for the diagonal magnetoconductivity reads [66]

$$\sigma_{xx}^{SdH}(\mu, H) = \frac{n_s e^2 \tau}{m^*} \frac{1}{1 + (\omega_{cyc}\tau)^2} \tag{3.10}$$

$$\times \left\{ 1 + 2\sum_{j=1}^{\infty}(-1)^j R\left(\frac{j2\pi r_{cyc}}{L_T}\right) \cos\left(j\frac{2\pi\mu}{\hbar\omega_{cyc}}\right) \exp\left(-j\frac{\pi}{\omega_{cyc}\tau}\right) \right\},$$

with n_s being the carrier density and m^* the effective mass of the charge carriers. For the simple case of cyclotron motion the velocity correlation function, action, etc. could be evaluated analytically.

σ_{xx}^{SdH} (see Fig. 3.4) represents the semiclassical approximation for the Shubnikov–de Haas (SdH) oscillations in terms of a sum over cyclotron orbits and their higher repetitions.

An expression corresponding to (3.8) exists also for the semiclassical Hall conductivity. However, for the case of a free 2DEG the two terms in the sum in (3.8) nearly cancel and second-order $\hbar$ corrections have to be considered to reproduce Hall plateaus in σ_{xy}^{SdH} and ρ_{xy}^{SdH} [169].

The expression (3.10) coincides with the quantum mechanical result for a constant scattering time τ [49]. In the case of a high-mobility 2DEG ($\omega_{cyc}\tau \gg 1$) with a pronounced Landau-level structure the use of an energy-dependent self-energy and scattering time is usually required. This leads, for example, to a quadratic dependence of the SdH oscillations on the density of states of each Landau level. However, for the antidot arrays to be considered below, the superlattice potential strongly mixes the Landau levels and smooths out the spectral density (Fig. 3.5). Thus the use of an energy-independent scattering time is justified.

3.1.2 Antidot Lattices: Experiments

As already illustrated in Fig. 1.1 in the introduction, a periodic array of nanometer-sized holes in a semiconductor heterostructure results in a periodic potential landscape for the two-dimensional electron gas (2DEG), i.e. an antidot superlattice. The electrons move at the Fermi energy in between the periodically arranged antidots. In these systems the elastic mean free path (MFP) l and the transport MFP l_T are both considerably larger than the lattice constant a. On the other hand, the Fermi wavelength λ_F is smaller than a. Hence semiclassical approximations for electron transport are justified.

The combination of the superlattice and a perpendicular magnetic field H leads to a number of peculiar effects,[2] of which we note a few:

(i) The diagonal magnetoresistivity $\rho_{xx}(H)$ exhibits pronounced peaks at moderate fields $H < 1$ T, contrary to the unpatterned 2DEG. This behavior is displayed in Fig. 3.1 for three antidot lattices differing in the ratio d/a of the effective antidot diameter d to the lattice constant. The number of resolved peaks in ρ_{xx} and steps in ρ_{xy} (not shown) depends critically on the ratio d/a. The wider the antidots, the fewer peaks are observable in the resistivity. The peaks were assigned to cyclotron orbits fitting around certain numbers of antidots [152]: for instance, for the sample 3 shown in Fig. 3.1 all peaks are located at field strengths such that the corresponding cyclotron radius $r_{\mathrm{cyc}} = cv_F/eH$ is commensurate with the lattice: peaks were resolved up to orbits encompassing 21 antidots.

(ii) Measurements of ρ_{xx} at low temperature ($T \sim 0.4$ K) displayed additional oscillations superimposed upon the low-H resistance anomalies [51], as already mentioned in the introduction. The corresponding data for an antidot lattice with $d/a \sim 0.5$ (sample 1 in Fig. 3.1) were shown in Fig. 1.2, Sect. 1.1. There, ρ_{xx} from both patterned and unpatterned sample segments is compared. In the unpatterned part, $1/H$-periodic Shubnikov–de Haas oscillations reflect the Landau energy spectrum. The oscillations in the antidot segment reveal quite different behavior: they are H-periodic with a period corresponding to the addition of approximately one flux quantum through the antidot unit cell. At 4.7 K, the quantum oscillations are smeared out, while the characteristic ρ_{xx} peak at $2r_{\mathrm{cyc}} = a$ persists.

This temperature sensitivity implies that the effects are of a quantum nature. However, they cannot be attributed to interfering electron waves traversing the whole antidot array, as discussed in previous work [166], since the samples are too large to maintain phase coherence. In the introduction it was shown that the observed periodicity of these modulations could be semiclassically described by quantized periodic orbits [51], assuming that ρ_{xx} reflects density-of-states oscillations. This picture will be refined in the

---

[2] For experimentally oriented reviews see, for example, [52, 164, 165].

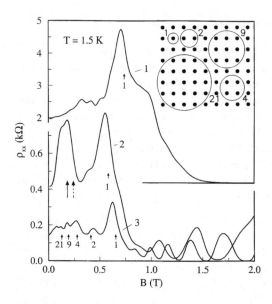

Fig. 3.1. Magnetoresistivity anomalies at low magnetic field for three samples differing in the ratio d/a of the antidot diameter to the lattice constant. For smaller d/a more structure in ρ_{xx} evolves. The peaks, e.g. in the curve 3, appear exactly at field strengths which correspond to the cyclotron radii r_{cyc} of commensurate orbits around 1, 2, 4, 9, and 21 antidots, as sketched in the *inset*. The corresponding r_{cyc}/a values are marked by *arrows*. The occurrence of the resistivity peaks is explained in Sect. 3.1.1. The *dashed arrow* at curve 2 marks the position of an unperturbed cyclotron orbit around four antidots. The shift of the corresponding resistance peak towards lower H indicates the deformation of the orbit in a "soft" potential [170] (from [152], by permission; ©1991 by the American Physical Society)

following section, where we present a uniform treatment of the two characteristic transport anomalies in antidot crystals: the broad *classical* enhancement of the magnetoresistance and the *quantum* oscillations.

3.1.3 Antidot Lattices: Conductivity Calculations

The transport anomalies in antidot lattices can be explained both within a numerical quantum mechanical approach [176, 177] and within the semiclassical framework presented above. We shall take the semiclassical route in the following and refer to the related quantum approaches afterwards.

Classical Dynamics

Fleischmann, Geisel, and Ketzmerick [170] were able to explain the *classical* magnetotransport anomalies observed, namely the strong peaks shown in Fig. 3.1, by accounting for the nonlinear classical electron dynamics in antidot lattices. As a classical model Hamiltonian for the motion of electrons at the Fermi energy under the effect of a magnetic field and the antidot lattice, they introduced

$$\mathcal{H} = \mu = \frac{1}{2m^*} \left(\boldsymbol{p} - \frac{e}{c} \boldsymbol{A} \right)^2 + U_0 \left[\sin \left(\frac{\pi x}{a} \right) \sin \left(\frac{\pi y}{a} \right) \right]^\beta . \qquad (3.11)$$

Here, the vector potential $\boldsymbol{A}$ generates the magnetic field and the parameter β governs the steepness of the antidots. Within a classical Kubo-type approach, Fleischmann et al. used for the conductivity

$$\sigma_{ij}^{\text{cl}} = p_c \frac{n_s e^2}{k_B T} \int_0^\infty \exp\left(-t/\tau\right) \langle v_i(t) v_j(0) \rangle_{\boldsymbol{pr}} \, dt \quad , \qquad (3.12)$$

where the angle brackets denote averaging over phase space. $0 < p_c < 1$ denotes the (chaotic) part of phase space in which particles can participate in transport. For an isotropic system the magnetoresistivity tensor is then obtained by means of

$$\rho_{xx} = \frac{\sigma_{xx}}{\sigma_{xx}^2 + \sigma_{xy}^2} \quad , \quad \rho_{xy} = \frac{\sigma_{xy}}{\sigma_{xx}^2 + \sigma_{xy}^2} . \qquad (3.13)$$

A corresponding numerical classical simulation [170] indeed reproduces most of the features of the classical magnetoanomalies shown in Fig. 3.1. The prominent resistivity peaks for $r_{\text{cyc}} \simeq a/2$ can be related to electrons temporarily trapped on orbits which encompass an antidot in an irregular manner. This motion leads to structure in the classical velocity correlation functions which is reflected in the resistivity.[3]

Indeed, (3.12) corresponds precisely to the smooth classical part (3.5) of the magetoconductivity obtained within the semiclassical approach. This is expected as the classical linear-response treatment should be included as the classical limit in the general semiclassical framework. For low temperatures the prefactor $n_s e^2 / k_B T$ in (3.12) has only to be replaced by $e^2 \bar{d}/V$.

Quantum Corrections

In the following we address the quantum oscillations observed in antidot experiments at low temperature (see Fig. 1.2). We shall show that they just reflect leading order $\hbar$ quantum corrections to the conductivity given by the semiclassical trace formula (3.6). A correct application of such trace formulas requires a detailed knowledge of the classical phase space structure. This is displayed by means of Poincaré surfaces of section in Fig. 3.2.

The Y coordinate in the figure denotes the diagonal (11) direction in a repeated unit cell of the antidot lattice (see, e.g., Fig. 3.3). The surface-of-section plots monitor the Y position and the velocity in the Y direction each time an orbit hits the diagonal. The three panels show the phase space

---

[3] Also, trajectories "hopping" through the array, sometimes denoted as *runaway* trajectories, can contribute to an increased resistivity; see [52, 154, 171, 172].

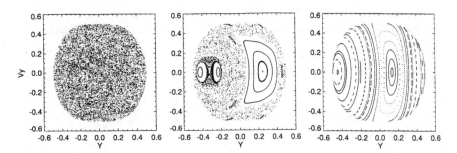

Fig. 3.2. Poincaré surfaces of section for the antidot model Hamiltonian (3.11) for $\beta = 2$ and field strengths $H = 0.5$, 1.5, and 4 T, from the *left* to the *right panel.* The normalized width of the antidots is $d/a = 0.5$. The surface of section is defined by the diagonal, the (11) direction in the lattice

structure of the Hamiltonian (3.11) for three different magnetic field strengths $H = 0.5, 2$, and 4 T and a steepness parameter $\beta = 2$. V_0 is determined through the normalized width d/a of the antidots at the Fermi energy. Here we use $d/a = 0.5$. At large H the classical motion is dominated by the magnetic field. The dynamics is quasi-regular and organized on tori. It turns out that in the regime below $H = 1.5$, where the quantum oscillations are observed, most of the classical phase space is chaotic, with a few embedded islands of stable motion which vanish with decreasing magnetic field. Below ~ 1 T (left panel in Fig. 3.2) the dynamics is practically completely chaotic, i.e. most of the periodic orbits are unstable.

In order to calculate $\sigma_{xx}^{\mathrm{osc}}(H)$ numerically it is convenient to expand the prefactors in (3.6) into geometrical series and to perform the sum over j to obtain the following form [66]:[4]

$$\sigma_{xx}^{\mathrm{osc}}(\mu, H) = \frac{4g_s}{V}\frac{e^2}{h}\,\mathrm{Re}\left\{\sum_{\mathrm{po}}\mathcal{C}_{xx}^{\mathrm{po}}\frac{\tau_{\mathrm{po}}}{\tau_T}\sum_{k,l=0}^{\infty}\frac{t_{\mathrm{po}}^{(k,l)}}{\left(1 - t_{\mathrm{po}}^{(k,l)}\right)^2}\right\} \qquad (3.14)$$

with

$$t_{\mathrm{po}}^{(k,l)} = (\pm 1)^k\,\exp\left[\mathrm{i}\left(\frac{S_{\mathrm{po}}}{\hbar} - \frac{\pi}{2}\eta_{\mathrm{po}}\right)\right]$$

$$\times \exp\left[-\left(k + \frac{1}{2}\right)\lambda_{\mathrm{po}} - (2l + 1)\frac{\tau_{\mathrm{po}}}{\tau_T} - \frac{\tau_{\mathrm{po}}}{2\tau}\right]. \qquad (3.15)$$

Here, $\lambda_{\mathrm{po}} = \mathrm{i}\gamma_{\mathrm{po}}$, with γ_{po} real, is the winding number and k a semiclassical quantum number for a stable orbit. λ_{po} is the Liapunov exponent in the case

---

[4] This resummed trace formula allows for a consistent summation of periodic-orbit contributions up to a given length. Such a representation is appropriate for applying periodic-orbit summation techniques like the cycle expansion in a regime where the trace formula is not absolutely convergent; see e.g. [35].

of an unstable orbit ($\lambda_{\mathrm{po}} > 0$ and real). The "$-$" sign in (3.15) applies only to unstable inverse hyperbolic orbits.

Fig. 3.3. Fundamental periodic orbits in a model antidot potential (3.11) which enter into the numerical evaluation of the trace formula (3.14)

In the following we compare the calculations with the experimental results. The experimental data for σ_{xx}, shown in the top panel of Fig. 3.4, were obtained by inverting the resistivity tensor (3.13) and by subtracting the nonoscillatory part of the conductivity, taken at higher temperatures. Under the experimental conditions ($T = 0.4$ K, $\omega_{\mathrm{cyc}}\tau \approx 2$ at $2r_{\mathrm{cyc}} \simeq a$ [51]) only the shortest periodic orbits contribute significantly to $\sigma_{xx}^{\mathrm{osc}}$ since the terms from longer orbits ($\tau_{\mathrm{po}} > 2\tau$ or $\tau_{\mathrm{po}} > \tau_T$) are exponentially small. These damping mechanisms select a limited number of relevant orbits which suffice to quantitatively describe the quantum oscillations. For the calculations the fundamental orbits shown in Fig. 3.3 were considered. Their stability exponents, actions, periods, and velocity correlation functions were numerically obtained for the model Hamiltonian (3.11).

The result for $\sigma_{xx}^{\mathrm{osc}}$ (at 0.4 K) as a function of magnetic field is shown as the solid line in the middle panel of Fig. 3.4: $\sigma_{xx}^{\mathrm{osc}}$ oscillates with the same frequency as the measured diagonal conductivity. The period of the oscillations is nearly *constant* with respect to H, in contrast to the $1/H$-periodic behavior of ordinary Shubnikov–de Haas oscillations (see (3.10)). The latter are shown in bottom part of Fig. 3.4 for an unmodulated 2DEG under the same conditions. In the light of the semiclassical trace formula, the H-periodic behavior in certain antidot arrays has a simple physical origin: each periodic orbit causes a modulation of $\sigma_{xx}(\mu, H)$ with a phase given by its H-dependent action

$$S_{\mathrm{po}}(H) = \oint \left(m^* \boldsymbol{v} + \frac{e}{c} \boldsymbol{A} \right) \mathrm{d}\boldsymbol{r} = m^* \oint \boldsymbol{v} \, \mathrm{d}\boldsymbol{r} - \frac{e}{c} H \mathcal{A}_{\mathrm{po}}(H) \,, \qquad (3.16)$$

where $\mathcal{A}_{\mathrm{po}}(H)$ denotes the enclosed area. With decreasing H, periodic orbits in certain antidot potentials cannot expand as is the case for free cyclotron

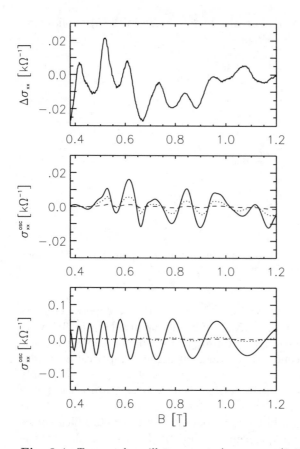

Fig. 3.4. *Top panel:* oscillatory part $\Delta\sigma_{xx} = \sigma_{xx}(0.4K) - \sigma_{xx}(4.7K)$ of the magnetoconductivity σ_{xx} (see (3.13)) of an antidot array from experiment [51]. *Intermediate panel:* semiclassically calculated $\sigma_{xx}^{\mathrm{osc}}$, (3.14), for $T = 0.4$ K (*solid line*), 2.5 K (*dotted line*), and 4.7 K (*dashed line*). *Bottom panel:* Shubnikov–de Haas oscillations, (3.10), for an unmodulated 2DEG under the same conditions as for the antidot lattice. (From [66], by permission)

orbits. Hence, $\mathcal{A}_{\mathrm{po}}(H)$ remains constant (approximately of the order of the unit cell for the fundamental orbits), and the magnetic field enters linearly into the semiclassical phase [51]. This behavior is characteristic of "soft", broad antidot potentials, while in lattices with more space between the antidots the oscillations are, rather, $1/H$-periodic, again in agreement with the corresponding semiclassical approximation [51, 173].

The semiclassical curves in the middle panel of Fig. 3.4 also exhibit the correct magnitude of the quantum oscillations and a temperature dependence in accordance with experiment: owing to the damping $R(\tau_{\mathrm{po}}/\tau_T)$ the periodic-orbit oscillations decrease for $T = 2.5$ K (dotted line in the middle panel of Fig. 3.4) and nearly disappear at 4.7 K (dashed curve) as in the experiment.

Temperature is the main damping mechanism; the results do not depend strongly on the elastic scattering time τ [66].

We note that the amplitude of the Shubnikov–de Haas oscillations, representing the integrable case, is considerably larger than that of the oscillations from isolated periodic orbits in the antidot array exhibiting chaotic classical dynamics.[5] This reflects the difference of $\sqrt{\hbar}$ between the two-dimensional integrable and chaotic cases.

In summary, interference effects due to summation over different periodic orbits, resulting in an amplitude modulation of the σ_{xx} oscillations, are clearly manifested in both experimental (top panel of Fig. 3.4) and calculated traces (middle panel of Fig. 3.4) for an antidot array.

Quantum Mechanical Calculations

Quantum mechanical calculations of magnetotransport in antidot lattices are far from being an easy task. They are rather involved owing to the strong coupling between the lattice potential and the magnetic field in the presence of weak disorder. Different approaches have been proposed: the self-consistent Born approximation on the basis of Green functions [175] or appropriately chosen eigenstates of the magnetotranslation group [176, 177], the recursive Green function technique [178], and an S-matrix formalism [179], to name a few. Quantum transport in antidot structures has been recently reviewed by Ando et al. [180] and Suhrke and Rotter [181]. Figure 3.5a displays results from elaborate quantum calculations [176] of the energy spectrum for different directions in the magnetic Brillouin zone of a soft antidot potential. Figure 3.5b shows the corresponding density of states. The insertion of the antidots into the 2DEG leads to a complex magnetic band structure. The rich, spiky structure in the density of states at $T = 0$ (thin solid line in Fig. 3.5b) is smoothed out for finite temperature (thick solid line). H-periodic oscillations remain as predicted by the semiclassical trace formula for the density of states.

The quantum calculations account for both classical and oscillatory terms in the antidot anomalies of the conductivity tensor. For soft antidot potentials the anomalies are indeed found to be periodic in H [181]. However, we note that quantum calculations for *rectangular* lattices reveal a phase shift in the oscillatory part of σ_{xx} with respect to the oscillations in the density of states that cannot yet be accounted for in the present semiclassical picture [182]. Quantum mechanically, the phase shift can be ascribed to the different roles of band and scattering conductivity, a concept which is beyond today's semiclassical approaches. Related differences have been reported as a result of a direct comparison between semiclassical computations and numerical quantum calculations [175, 179]. These differences at small H were assigned to

---

[5] For a semiclassical treatment of integrable superlattices see e.g. [174].

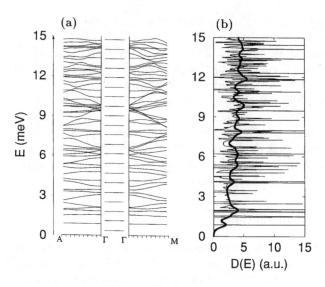

Fig. 3.5. Band structure (**a**) and corresponding density of states (**b**) of electrons in an antidot lattice calculated for $B = 0.41$ T, $a = 200$ nm, and $\beta = 2$ by Silberbauer [183]. The density of states is shown for $T = 0$ (*thin solid line*) and for $T = 1.5$ K (*thick solid line*)

possible couplings between the periodic orbits, mediated by nonclosed skipping orbits.

Furthermore, the use of a constant self-energy, as in the present semiclassical approach, is no longer justified for superlattices with pronounced band structure, i.e. variations in the density of states. Hence, a corresponding further refinement of the present semiclassical periodic-orbit transport theory is desirable.

3.2 Transport Through Phase-Coherent Conductors

The transport phenomena of the "macroscopic" superlattices discussed above reflect a bulk-like, though peculiar, behavior since the size of the devices exceeds the phase-coherence length ℓ_ϕ considerably. Therefore, it was hoped to detect specific signatures of phase-coherent transport in antidot lattices with dimensions smaller than ℓ_ϕ [144, 185]. A corresponding measurement in which a finite array of antidots was placed in a micron-sized two-dimensional cavity [144] displayed resistance fluctuations very similar to those first observed by Marcus et al. [55] in transport experiments through billiard-like nanostructures of different geometries. These and related experiments have addressed the question of interference in phase-coherent ballistic nanostructures giving rise to conductance fluctuations [55, 141–143] and quantum corrections to the average conductance [44, 145, 148].

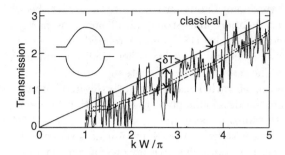

Fig. 3.6. Total transmission coefficient (as a function of wave vector k, in units of the mode number in the leads of width W) corresponding to the conductance through a phase-coherent ballistic cavity. The *fluctuating line* is the full quantum transmission T_{qm}, and the *straight solid line* is the classical transmission T_{cl}. The *dashed* and *dotted lines* denote the averaged transmission T_{qm} at small magnetic field $H = 0$ and $HA/\phi_0 = 0.25$. The smoothed T_{qm} are substantially smaller than T_{cl} by an H-dependent amount $\langle \delta T \rangle$ (from [41], by permission)

The main features of ballistic transport through microstructures are summarized in Fig. 3.6 [41, 43]. The figure displays results for the numerically calculated total transmission coefficient for an asymmetric stadium cavity with two leads of width W attached. The following effects are visible:

(i) The classical transmitted flux, obtained from the shooting in an appropriate distribution [186] of classical particles and counting the transmitted ones, increases linearly as a function of wave number.

(ii) The quantum transmission exhibits strong oscillations with a size of order unity: ballistic conductance fluctuations.

(iii) The averaged quantum transmission increases roughly linearly with the same slope as the classical transmission, but lies below the classical line. This is predominantly due to mode effects from confinement in the leads; note the stepwise increase of the quantum transmission on the scale of kW/π. However, part of the offset is sensitive to a weak magnetic field H. The reduction of quantum transmission is strongest for zero field. This illustrates the weak-localization effect.

The main underlying picture, to be developed in the following, is that quantum effects in ballistic transport as depicted in Fig. 3.6 can be ascribed to interference among complicated boundary-reflected paths in the cavity.

3.2.1 Semiclassical Approach to Landauer Conductance

A semiclassical approach represents the natural theoretical tool to incorporate this picture into a theory of the transmission. The Landauer formalism serves as an appropriate general framework to deal with phase-coherent transport and to provide a link between transmission and conductance [10, 187].

A corresponding semiclassical version of Landauer transport theory has been developed mainly by Baranger, Jalabert and Stone [42, 43, 56]. It is partly based on work by Blümel and Smilansky [188] on the related problem of chaotic scattering and on early pioneering work by Miller [259] introducing semiclassical reaction theory.[6] The semiclassical concepts introduced for ballistic quantum transport have proven rather useful for both the understanding of related experiments and the development of the connection between classical and quantum properties for open systems.

We only briefly sketch the main steps in the derivation of a semiclassical formula for the conductance. For comprehensive reviews of semiclassical aspects of ballistic quantum transport see [41, 43]. To illustrate the semiclassical approach, we consider ballistic conductance fluctuations and weak localization. As a specific application we present a comparison with transport experiments on finite antidot arrays, making contact with the preceding section. The treatment of the weak-localization effect in terms of coherent backscattering of paths shows the advantages but also the limitations of present semiclassical transport theory, which will be addressed in Sect. 3.3.

According to the Landauer formalism the conductance G through a phase-coherent sample attached to two leads is simply proportional to the total transmission T at energy E. For leads of width W that support N current-carrying modes the conductance reads

$$G = g_s \frac{e^2}{h} T = g_s \frac{e^2}{h} \sum_{n,m=1}^{N} |t_{nm}|^2 \, . \tag{3.17}$$

The $t_{nm}(E)$ denote transmission amplitudes between incoming channels m and outgoing channels n in the leads. Without backscattering the conductance increases by $g_s e^2/h$ each time a new mode opens. This effect is still apparent in the conductance of more complicated scatterers, as the residual staircase structure of the quantum transmission in Fig. 3.6 shows.

The amplitudes t_{nm} in (3.17) can be written in terms of the projections of the Green function onto the transverse modes $\phi_n(y')$, $\phi_m(y)$ in the leads (at x and x', respectively):

$$t_{nm} = -i\hbar (v_n v_m)^{1/2} \int dy' \int dy \, \phi_n^*(y') \phi_m(y) \, G(x', y', x, y; E) \, , \tag{3.18}$$

where the v_n denote longitudinal velocities. The integrals are taken over the cross sections of the (straight) leads at the entrance and the exit.

The semiclassical approximation enters on two levels: first, by replacing $G(x', y', x, y; E)$ in (3.18) by the semiclassical Green function (2.3) in terms of classical paths; second, by evaluating the projection integrals for isolated trajectories within the stationary-phase approximation. For leads with hard-wall boundaries the mode wavefunctions are sinusoidal, $\phi_m(y) =$

---

[6] This theory, developed for atom-exchange reactions in molecular scattering, exhibits a close formal relationship to the Landauer formalism.

$\sqrt{2/W}\sin(m\pi y/W)$. Hence the stationary-phase condition for the y integral requires

$$\left(\frac{\partial S}{\partial y}\right)_{y'} = -p_y \equiv -\frac{\overline{m}\hbar\pi}{W} , \qquad (3.19)$$

with $\overline{m} = \pm m$. The stationary-phase solution of the y' integral yields a corresponding "quantization" condition for the transverse momentum $p_{y'}$. Thus only those paths which enter into the cavity at (x, y) with a fixed angle $\sin\theta = \pm m\pi/kW$ and exit the cavity at (x', y') with angle $\sin\theta' = \pm n\pi/kW$ contribute to $t_{nm}(E)$. There is an intuitive explanation: the trajectories are those whose transverse wave vectors on entrance and exit match the wave vectors of the modes in the leads.

One then obtains for the semiclassical transmission amplitudes

$$t_{nm} = -\frac{\sqrt{2\pi \mathrm{i}\hbar}}{2W} \sum_{t(\overline{n},\overline{m})} \mathrm{sgn}(\overline{n})\,\mathrm{sgn}(\overline{m})\sqrt{A_t}\exp\left[\frac{\mathrm{i}}{\hbar}\tilde{S}_t(\overline{n},\overline{m};k) - \mathrm{i}\frac{\pi}{2}\tilde{\mu}_t\right]. \quad (3.20)$$

Here, the reduced actions are

$$\tilde{S}_t(\overline{n},\overline{m};k) = S_t(k) + \hbar k y \sin\theta - \hbar k y' \sin\theta' \qquad (3.21)$$

$(k = \sqrt{2mE}/\hbar)$, which can be considered as Legendre transforms of the original action functional. The phases $\tilde{\mu}_t$ contain both the usual Morse indices and additional phases arising from the y, y' integrations. The prefactors are $A_t = |\partial y/\partial\theta'|/(W\cos\theta')$. The resulting semiclassical expression for the transmission and thereby the conductance (see (3.17)) in chaotic cavities involves contributions from pairs of trajectories t, t'. It reads [42, 43]

$$T(k) = \sum_{n,m=1}^{N} |t_{nm}(k)|^2 = \frac{\pi}{2kW} \sum_{n,m=1}^{N} \sum_{t,t'} F_{n,m}^{t,t'}(k) , \qquad (3.22)$$

with

$$F_{n,m}^{t,t'}(k) \equiv \sqrt{A_t A_{t'}}\,\exp\left[\frac{\mathrm{i}}{\hbar}(\tilde{S}_t - \tilde{S}_{t'}) + \mathrm{i}\mu_{t,t'}\frac{\pi}{4}\right]. \qquad (3.23)$$

The phases $\mu_{t,t'}$ account for the differences in the phases $\tilde{\mu}_t$ and the sgn factors in (3.20).

Equations (3.20), (3.22), and (3.23) may serve as a general starting point in the following ways:

(i) as a semiclassical tool to analyze quantum mechanical transmission;
(ii) for a direct evaluation of the quantum conductance by summing the different path contributions numerically;
(iii) for further approximations which enable the derivation of analytical semiclassical predictions, e.g. for weak localization and correlations in the conductance.

3.2.2 Trajectory Analysis

We first proceed along the lines of (i) to extract hidden physical information from experimentally or numerically obtained complex transmission spectra such as that of Fig. 3.6 Usually a direct comparison of the strongly fluctuating quantum and semiclassical transition amplitudes shows only little correspondence [189]. The form of the semiclassical expression (3.20) for the transmission amplitude, however, suggests that a Fourier transform of $t_{nm}(k)$ should yield a power spectrum with peaks at the lengths L_t of the lead-connecting trajectories, since the actions in billiards scale linearly with wave vector, $S_t(k) = kL_t$ (at $H = 0$). Such a Fourier analysis has been carried out in [189, 191] for the case of a circular billiard and in [192] for a circular billiard with a tunnel barrier.

A result of a comprehensive analysis by Delos and coworkers [189] is displayed in Fig. 3.7. In the right panel the power spectra of the numerically calculated quantum transmission amplitude $t_{11}(k)$ and of the corresponding semiclassical transmission amplitude (based on 120 trajectories) are compared. The agreement between the quantal and semiclassical results is remarkable up to scaled lengths $L \sim 20$. The lengths L_t of the orbits, shown as insets above some of the peaks, precisely mark the locations of the peaks on the length axis. To obtain the peak heights to such an accuracy, Delos et al. did not employ the stationary-phase approximation for the integrals over the cross sections, which becomes questionable for the low channel numbers $n = m = 1$ used; instead they accounted for diffraction effects at the lead apertures. This amounts to treating the initial wave front as circular and to launching trajectories from the center of the lead mouth in all directions.

We note that a semiclassical analysis (e.g. [189, 191, 192]) and synthesis [190]) of transition *probabilities* $|t_{nm}|^2$ instead of *amplitudes* is more involved since one has to deal with orbit pairs and thus phases $k(L_t - L_{t'})$ arising from their length differences in the Fourier transform.

3.2.3 Weak Localization

In the following we evaluate the expression (3.22) using further approximations to derive semiclassical estimates of statistical properties of conductance oscillations and of the coherent backscattering from time-reversed paths. This mechanism contributes to the quantum enhancement of the average magnetoresistance near $H=0$ known as the weak-localization effect.

The semiclassical evaluation of the reflection coefficient $R = N - T$ proceeds in much the same way as the derivation of T, but considers orbits being backscattered to the entrance lead. The classical reflection coefficient is obtained, as in the Kubo formalism, by pairing identical paths $t = t'$. Quantum corrections δR to R are conveniently decomposed into a part diagonal in the mode index and nondiagonal terms according to [43, 56]:

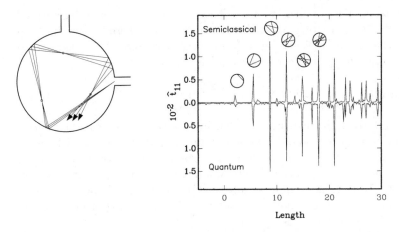

Fig. 3.7. Comparison of semiclassical and numerical quantum results for the power spectrum of the transmission amplitude $t_{11}(k)$ for a circular geometry as shown in the *left panel*. Each major peak corresponds to one classical path (shown in the *insets*) between entrance and exit lead. (From [189], by permission; ©1996 by the American Physical Society)

$$\delta R = \delta R_{\mathrm{D}} + \delta R_{\mathrm{ND}} \equiv \frac{\pi}{2kW} \left\{ \sum_{n}^{N} \sum_{t \neq t'} F_{n,n}^{t,t'} + \sum_{n \neq m}^{N} \sum_{t \neq t'} F_{n,m}^{t,t'} \right\} . \qquad (3.24)$$

An approximate analytical evaluation can be given for the low-field dependence of the ensemble-averaged diagonal part, $\langle \delta R_{\mathrm{D}} \rangle$. To this end we assume that off-diagonal pairs of paths $t \neq t'$ with differences in their dynamical phases $ik(L_t - L_{t'}) \neq 0$ average out. Only pairs consisting of a path and its time-reversed partner are assumed to survive an ensemble (energy) average, since their dynamical phases cancel.

This represents the semiclassical diagonal approximation.[7] In the presence of a small magnetic field these pairs retain a flux-dependent phase due to the opposite signs of the enclosed flux. This phase is given by $(S_t - S_{t'})/\hbar = 2\Theta H/\phi_0$, where $\Theta = 2\pi \int_t \boldsymbol{A} \cdot \mathrm{d}\boldsymbol{l}/H$ is the directed enclosed "area" of a backscattered orbit and $\phi_0 = hc/e$ denotes the flux quantum.

In the semiclassical limit, the sum over the N reflection coefficients in (3.24) may be converted to an integral over an angle, namely $\pi/(kW) \sum_n \to \int \mathrm{d}(\sin\theta)$, and we obtain

$$\langle \delta R_{\mathrm{D}}(H) \rangle = \frac{1}{2} \int_{-1}^{1} \mathrm{d}(\sin\theta) \sum_{t(\theta,\theta),t(\theta,-\theta)} A_t \exp\left(\frac{\mathrm{i}2\Theta_t H}{\phi_0} \right) . \qquad (3.25)$$

Moreover, assuming that the escape time is larger than the time required to uniformly cover the phase space of the chaotic scatterer, the distribution of

---

[7] The diagonal approximation will be critically discussed in Sect. 3.3.

outgoing particles is uniform in $\sin \theta'$. Then the integral over backscattered paths contributing to $\langle \delta R_D \rangle$ can be replaced by an average over injected particles (the prefactor $|A_t|$ acts as a Jacobian). Regrouping the backscattered paths according to their effective area yields [42,56][8]

$$\langle \delta R_D(H) \rangle \sim \int_{-\infty}^{\infty} d\Theta \, N(\Theta) \cos\left(\frac{2\Theta H}{\phi_0}\right) \,. \tag{3.26}$$

Here $N(\Theta)$ denotes the distribution of enclosed areas of backscattered paths. Hence, in the semiclassical diagonal approximation, the coherent-backscattering term of the average resistance is given by the cosine transform of the corresponding area distribution function, a purely classical quantity. Assuming that $N(\Theta)$ decreases exponentially, $N(\Theta) \sim \exp(-\alpha_{cl}\Theta)$, leads directly to the semiclassical prediction of a *universal* Lorentzian H dependence of $\langle \delta R_D(H) \rangle$ for *chaotic* systems [42,56]:

$$\langle \delta R_D(H) \rangle \sim \frac{\mathcal{R}}{1 + (2H/\alpha_{cl}\phi_0)^2} \,. \tag{3.27}$$

$\mathcal{R}$ is the classical reflection coefficient. The width of the Lorentzian is system-specific. It depends via

$$\alpha_{cl} \sim \sqrt{\gamma_{cl}W}/\Theta_0 \tag{3.28}$$

on the parameter Θ_0, the typical area per circulation, and the classical escape rate γ_{cl} [194] for trajectories to leave the cavity.

In contrast, the line shape of the average magnetoresistance for integrable cavities is expected to be nonuniversal, in particular non-Lorentzian-like, because the area distribution typically decays with a system-specific power law. This difference between regular and hyperbolic behavior, originally predicted semiclassically in [56], has indeed been observed in experiments [44,145] when comparing the magnetoconductance of an ensemble of ballistic stadia, supposed to be chaotic, with that of a circular microstructure. This was shown in Fig. 1.4 in the introduction. Further numerical quantum calculations have, on the whole, confirmed these findings [184].

3.2.4 Finite Antidot Arrays

As a further application of this semiclassical approach we consider weak-localization effects in quantum dots with internal structure: arrays of antidot scatterers arranged in square cavities.

An electron micrograph of such a finite antidot array, fabricated by electron beam lithography and dry etching techniques, is displayed in Fig. 3.8. The transport mean free path[9] l_T is ~ 16 µm, which is considerably longer

---

[8] A similar analysis has been performed in [193] in the context of microwave cavities.

[9] For a discussion of the transport mean free path see Appendix A.3 and Chap. 5.

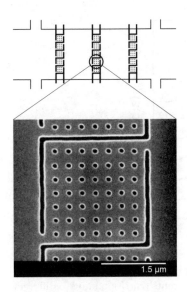

Fig. 3.8. Electron micrograph depicting one square-like microcavity containing 49 antidots with a lattice period $a \simeq 300$ nm. The transport mean free path of the unpatterned device is larger by a factor of approximately 7 than the cavity length L. The resistance is measured across an ensemble of cavities (*top*). (From [68], by permission)

than the length of the squares, $L \simeq 2.3$ µm. In order to suppress conductance fluctuations, which are observed for single antidot cavities, and to enable the unperturbed observation of the weak-localization peak, the averaged resistance of an arrangement of 52 cavities was measured [68]. The arrangement is shown schematically in Fig. 3.8.

The left-hand side of Fig. 3.9 displays characteristic averaged resistance peaks at small fields, centered at $H = 0$. The phase coherence length ℓ_ϕ was decreased (from top to bottom for the curves shown) by successive increases of the bias current, which leads to current heating of the electrons. The line shape of the weak-localization peak evolves correspondingly from a cusp-like peak at the lowest temperature to a Lorentzian-like profile in the limit of shorter ℓ_ϕ. This strong deviation from a Lorentzian line shape was rather unexpected in view of the semiclassical prediction for chaotic cavities given above.

The quantum mechanical properties of a finite array of antidot potentials in a cavity are rather complex owing to the interplay between the remnants of the band structure of the antidot lattice, boundary effects, and an external magnetic field [196]. This is reflected in the level diagram for the spectral density of the corresponding closed system, shown in Fig. 3.10. At small fields, signatures of a band structure are still present, while precursors of Landau levels are visible at strong fields.

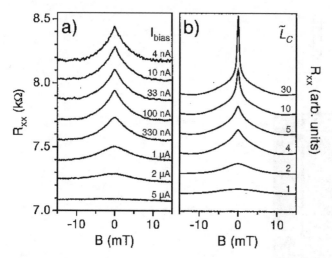

Fig. 3.9. (a) Measured magnetoresistance of an ensemble of square cavities containing antidots (as in Fig. 3.8) for different bias currents (corresponding to different temperatures). (b) Results from semiclassical simulations of quantum corrections to the resistance, (3.26), as a function of the cutoff length $\tilde{L}_c = L_c/L$ (corresponding to ℓ_ϕ). An offset has been added for clarity (adapted from [68], by permission)

Thus, to understand the experimental results at least qualitatively, the use of semiclassical concepts seems appropriate. The averaged reflection coefficient was calculated according to (3.24) [68]. The area distribution $N(\Theta)$ was numerically obtained by injecting $\sim 10^6$ electrons into a cavity filled with antidots as shown in Fig. 3.11a. The model potential (3.11) was used as the antidot potential. In the numerical simulations a cutoff length L_c was introduced as a crude means to account for the effect of the temperature-dependent phase breaking. L_c defines the maximum length of paths contributing to $N(\Theta)$. Figure 3.9b displays the evolution of the computed peak profile on decreasing the normalized cutoff length $\tilde{L}_c = L_c/L$ from 30 to 1. One finds a crossover from a cusp-like peak to a smooth, flat profile, very similar to that observed in the experiment. By relating the simulations to the experimental traces one can get a rough estimate of the phase-coherence length of $\ell_\phi \approx 5L \simeq 12$ µm at the lowest temperature.

The deviation from the Lorentzian peak profile expected for classically chaotic systems is due to the coexistence of two types of trajectories, shown in Fig. 3.11a: short paths of electrons directly reflected by the antidots near to the cavity entrance or by the opposite cavity wall, and longer trajectories where the electrons move diffusively through the array of scatterers before leaving the cavity. Directly backscattered paths enclose small areas and thus contribute to the area distribution $N(\Theta)$ as a peak at small Θ, which translates into a *broad* weak-localization peak, shown in the inset of Fig. 3.11. On the other hand, the long chaotic trajectories show an exponential decrease of

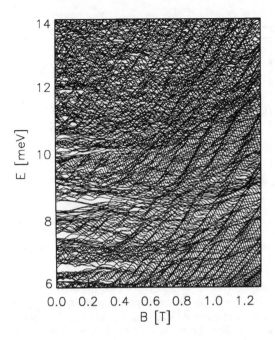

Fig. 3.10. Quantum mechanically calculated energy level diagram for an array of antidots arranged in a closed square cavity as a function of an applied perpendicular uniform magnetic field. The energy scale corresponds to that of the related experiment. A complex spectral structure results from the competing effects of confinement, lattice potential, and magnetic field

$N(\Theta)$ in accordance with chaotic scattering theory. They give rise to a *narrow* resistance cusp, being very sensitive to changes in ℓ_ϕ. In summary, one can conclude that in a typical chaotic system quantum transport phenomena usually reflect the coexistence of short-time dynamics of nonuniversal paths *and* chaotic behavior on longer timescales. Therefore the generic chaotic feature of a pure Lorentzian weak-localization peak appears to be quite exceptional. In general the resistance profile is expected to be more complex and not unique.

3.2.5 Conductance Fluctuations

Other than the weak-localization effect, conductance fluctuations represent the most striking quantum features of ballistic transport (see Fig. 3.6). These irregular-looking oscillations are defined by the deviation of the quantum conductance T from its average part, $\delta T = T - \mathcal{T}kW/\pi$, where $\mathcal{T}$ is the classical transmission (an asymmetrized system is assumed). δT as a function of magnetic field or wave vector is conveniently characterized in terms of correlation functions [43]:

$$C(\Delta H) \equiv \langle \delta T(k, H + \Delta H) \delta T(k, H) \rangle_k , \qquad (3.29a)$$

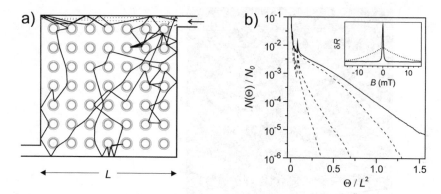

Fig. 3.11. (a) Model of a 7×7 antidot cavity. In the classical simulation electrons injected from the *right lead* are either directly backscattered by nearby antidots (*dashed line*) or the opposite boundary (*dotted lines*), or follow paths in a diffusive way through the array of scatterers (*solid line*). (b) Classical area distribution $N(\Theta)/N_0$ for backscattered paths for the antidot "billiard" (*solid line*). The nearly exponentially decreasing tail at larger directed areas Θ represents diffusive electron motion. It is suppressed upon reduction of the cutoff length $\tilde{L}_c$ (*dashed lines* for $\tilde{L}_c \approx 30, 5, 2$). *Inset*: decomposition of the weak-localization peak (see Fig. 3.9b) into contributions from long chaotic orbits (Lorentzian-like spike, *solid line*) and orbits shorter than $3L$ (*dotted line*) (from [68], by permission)

$$C(\Delta k) \equiv \langle \delta T(k + \Delta k, H) \delta T(k, H) \rangle_k \; . \tag{3.29b}$$

The semiclassical technique for evaluating these correlation functions follows closely the approaches by Gaspard and Rice [194] and Blümel and Smilansky for S-matrix correlations [188]. It is similar in spirit to the semiclassical treatment of weak localization in Sect. 3.2.3 and is reviewed in detail in [41, 43]. Therefore we only summarize the main results. The correlation functions contain products of δT and thus, semiclassically, contributions from four sets of classical paths. Again wave vector averaging is performed within the diagonal approximation by pairing those trajectories where the sum of all phase factors is canceled except those arising from differences ΔH or Δk. As a result one finds the following approximate correlation functions for ballistic conductance fluctuations of *chaotic* systems [41, 141]:

$$C_D(\Delta H) \sim \left[\frac{1}{1 + (\Delta H / \alpha_{\text{cl}} \phi_0)^2} \right]^2 , \tag{3.30a}$$

$$C_D(\Delta k) \sim \frac{1}{1 + (\Delta k / \gamma_{\text{cl}})^2} \; . \tag{3.30b}$$

α_{cl} was defined in (3.28) and γ_{cl} is the classical escape rate. While the agreement with numerical quantum calculations is good [43], the agreement with experimental results on semiconductor billiards may be considered as only fair [55, 142].

The above results rely on the assumption that the scatterer is completely hyperbolic, while in many realizations the confinement potentials give rise instead to a mixed phase space structure where regular and chaotic regimes coexist. The classical escape from such cavities usually follows a power law, i.e. it is much slower than the exponential escape from an open chaotic system. The power-law decay originates from the "trapping" of chaotic classical orbits in the neighborhood of an (infinite) hierarchy of regular regions in phase space. Assuming a power law $P(t) \sim t^{-\beta}$ for the staying time in the cavity and $P(\Theta) \sim \Theta^{-\gamma}$ for the area distribution, the occurence of *fractal* conductance fluctuations, which reflect the hierarchical phase space morphology, was predicted [197]. The variance of conductance increments is expected to scale for small magnetic fields as $\sim (\Delta H)^{\gamma}$. To measure such fractal structures is experimentally challenging since it amounts to resolving conductance fluctuations on rather small scales. Moreover, a serious fractal analysis usually requires the observation of a power law over at least two orders of magnitude (in magnetic field). This has been recently achieved by Sachrajda et al. in an experiment on ballistic conductance fluctuations in a soft-wall semiconductor stadium [150].

In summary, semiclassics enables the study of transport for the three classes of ballistic conductors with hyperbolic, regular, and mixed classical dynamics.

3.3 Limitations of Present Semiclassical Transport Theory

After the presentation of the semiclassical approximations within the framework of the Kubo and Landauer transport formalisms in the preceding sections, a few remarks are due at this point. The approaches used above reflect the advantages and power as well as some of the remaining problems of semiclassical transport theory.

Semiclassical transport theory provides a simple and physically transparent tool to compute particular contributions to the full quantum conductance. In particular, the semiclassical approach provides a means to deal with systems that exhibit both regular and chaotic dynamics. Moreover, as shown for the finite antidot lattices, this method helps, at least qualitatively, to recover universal signatures of a chaotic structure in quantum transport which are masked by features from the short-time dynamics [191].

In this respect semiclassical theory exhibits advantages compared to random-matrix theory. The latter has been proven very powerful for dealing with transport phenomena in open chaotic systems [25, 198–201]. However, by its very nature, it cannot deal with transport through nonchaotic structures. Also, classical system-specific parameters like α_{cl} must be additionally included, for example to predict the correct scales of weak-localization peaks.

Since scattering systems generally exhibit coexisting regular and chaotic regimes in phase space, a further development of semiclassical transport theory is highly desirable.

With regard to the weak-localization effect, the semiclassical evaluation of the coherent backscattering contribution in the Landauer formalism yields characteristics of the average magnetotransport which only partly agree with experiment and numerical quantum calculations. Semiclassical theory, as discussed above, correctly gives the width and line shape of the weak-localization peak as a function of H. However, it fails in reproducing the correct peak height: the error is of the same order as the effect itself! This semiclassical approach relies on the diagonal approximation: only pairs of backscattered paths with their time-reversed partners are included in the first sum of the expression (3.24) for the averaged reflection coefficient. Moreover, it can be shown that other off-diagonal contributions to the magnetoresistance precisely cancel the semiclassical term arising from time-reversed backscattered paths [202].

The disappearance of the (leading-order!) quantum correction in this semiclassical approximation can be seen most directly when considering the conductance through a cavity in a form where the projection onto channel wavefunctions of the leads has not yet been performed. In terms of the nonlocal conductivity tensor $\boldsymbol{\sigma}(\boldsymbol{r}, \boldsymbol{r}')$ the conductance of a two-terminal geometry is given as [204]

$$G = -g_{\mathrm{s}} \int \mathrm{d}y' \int \mathrm{d}y \, \hat{\boldsymbol{x}} \, \boldsymbol{\sigma}(\boldsymbol{r}, \boldsymbol{r}') \, \hat{\boldsymbol{x}}' \,. \tag{3.31}$$

Here the integrals are over the two cross sections of the incoming and outgoing leads and $\hat{\boldsymbol{x}}$ and $\hat{\boldsymbol{x}}'$ are unit vectors normal to the cross sections. Expressing $\boldsymbol{\sigma}(\boldsymbol{r}, \boldsymbol{r}')$ through advanced and retarded Green functions, the conductance (at zero temperature) reads [204]

$$G = -\frac{g_{\mathrm{s}} e^2 \hbar^3}{8\pi m^2} \int \mathrm{d}y' \int \mathrm{d}y \, G^+(\boldsymbol{r}, \boldsymbol{r}'; E)(\overset{\leftrightarrow}{\boldsymbol{D}} \cdot \hat{\boldsymbol{x}})^*(\overset{\leftrightarrow}{\boldsymbol{D}} \cdot \hat{\boldsymbol{x}}')G^-(\boldsymbol{r}', \boldsymbol{r}; E). \tag{3.32}$$

Here $\overset{\leftrightarrow}{\boldsymbol{D}}$ is a double-sided derivative. If we replace the two Green functions by their semiclassical approximations and evaluate the two surface integrals by using stationary phase, similarly to the integrals in Appendix A.1, only those pairs of paths which lie on a periodic orbit contribute [203]. However, if the two surface integrals are taken over cross sections of two *infinite* leads connected to the conductor, the two conditions that the paths have to start and end at these cross sections and that they lie on a periodic orbit cannot be simultaneously fulfilled. Hence this argument leads to the same result as above: the absence of semiclassical contributions to quantum corrections to the conductance. The failure of semiclassics is linked to the fact that semiclassical transport theory in this approximation is not unitary. Unitarity implies that the total current flowing through the sample is independent of the locations of the two cross sections. They can also be put inside the cavity. However, with the argument based on (3.32), different contributions are obtained for different positions of the cross sections: a clear sign that current is not conserved semiclassically.

Nevertheless, despite these severe problems the qualitative agreement of the semiclassical magnetoresistance profile with experiment and numerical results suggests that the semiclassical coherent backscattering term reflects at least partly the underlying physical mechanisms.

A situation comparable to that sketched above exists for the semiclassical approximation within the framework of Kubo linear response theory (see Sect. 3.1). The derivation, as presented in Sect. 2.4, cannot account for quantum corrections to the average conductivity, i.e. weak localization in the ballistic regime. It is not surprising that both semiclassical approaches exhibit the same shortcomings: quantum mechanically, the Kubo and Landauer approaches to transport are equivalent [204]. Both complementary semiclassical derivations make use of the same approximations: Green functions are expressed in terms of classical paths and (trace) integrals are evaluated in the stationary-phase approximation.

This puzzle has led to the fundamental question of whether quantum weak localization can be explained at all by interfering purely classical paths. There exist suggestions that contributions from nonclassical paths possibly have to be included for an adequate description [41, 43]. A resolution of these open questions is still lacking. As argued by Argaman [128] and Aleiner and Larkin [205], the semiclassical stationary-phase evaluation is too crude: trajectories starting nearly but not completely parallel to each other have to be included in a more appropriate manner.

In order to understand this idea consider a ballistic system with antidots arranged on a regular lattice or at random. In this case standard weak-localization theory for disordered systems is not valid: this theory is well suited to describe coherent backscattering from (point-like) impurities where the scattering is regarded as a *quantum* process [7]. In particular, this provides a quantum mechanism for the "splitting" of classical trajectories at impurities, allowing the formation of pairs of flux-enclosing time-reversed backscattered paths. Moreover, the electronic motion can be regarded as a delta-correlated, diffusive process. Antidots with a diameter a considerably larger than the Fermi wavelength λ_F act as *classical* scatterers. Hence ballistic (antidot) systems call for a generalization of weak-localization theory beyond the diffusion approximation in order to account for correlations in the ballistic classical dynamics. According to [128] and [205], it is the exponential separation of initially close orbits in a chaotic system with classical scatterers which provides a mechanism for a minimal wave packet of size λ_F to split into two parts which then follow time-reversed paths before they interfere constructively upon return. This backscattering mechanism is sketched in Fig. 3.12 for a pair of returning paths with small differences in their initial phase space coordinates.

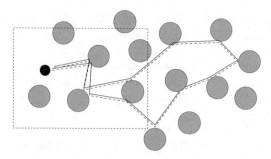

Fig. 3.12. Sketch of a pair of paths which contribute to coherent backscattering in a chaotic system with classical scatterers. Orbits of a minimal wave packet of size λ_F (*small filled circle*), which are initially close in phase space, separate on the scale of the Ehrenfest time (3.33) up to a distance of the order of the size a of the classical scatterers (Liapunov zone, marked as a *dashed rectangle*). This provides a mechanism for the initial wave packet to split into two parts which then follow time-reversed paths before they interfere constructively upon return

This approach introduces another relevant timescale for weak localization in a chaotic system: the Ehrenfest time [206],

$$t_E = \frac{1}{\lambda} \ln\left(\frac{a}{\lambda_F}\right) , \tag{3.33}$$

for the spreading of the wave packet over a distance of the order of the size a of the antidots. Here λ denotes the mean Liapunov exponent of the classical system. In line with this semiclassical picture Aleiner and Larkin have computed a weak-localization correction for ballistic systems. Their approach accounts for correlations in the chaotic ballistic dynamics in the "Liapunov region" (marked by dashed lines in Fig. 3.12) for timescales up to the Ehrenfest time. This is technically achieved within a ballistic σ model by replacing the diffusion operator by means of the regularized Liouville operator, the Perron–Frobenius operator. For times larger than t_E the classical mechanics is assumed to be uncorrelated and is treated as diffusive again. The result proposed for ballistic weak localization then reads [205]

$$\Delta\sigma = -\frac{e^2}{\pi h} \exp\left[-\frac{t_E}{\tau_\phi}\left(1 - \frac{\lambda_2}{\lambda^2 \tau_\phi}\right)\right] \ln\frac{\tau_\phi}{\tau}. \tag{3.34}$$

Here, τ_ϕ is the phase-breaking time and $\lambda_2 \sim \langle \delta\lambda(t_1)\delta\lambda(t_2)\rangle$ characterizes fluctuations in λ. In (3.34), correlations in the chaotic dynamics are incorporated into the exponential prefactor, while the diffusive motion on longer timescales is reflected in the logarithm. The standard weak-localization result for a two-dimensional diffusive system is recovered by setting t_E to zero.

A recently observed exponential temperature dependence of weak localization measured in periodic and irregular arrays of antidots is indeed in line with an analysis on the basis of (3.34) if one accounts for the temperature dependence of τ_ϕ (due to electron–electron and electron–phonon scattering) [156].

The unusual temperature behavior observed shows signatures of chaotic dynamics in weak localization. Interestingly, (3.34) may offer the possibility to extract the Ehrenfest time and, in view of (3.33), the *classical* Liapunov exponent of an electronic billiard from the *quantum* weak-localization correction.

The semiclassical picture behind the approach of Aleiner and Larkin is certainly physically appealing; the method of evaluating the result (3.34), however, makes use of semiclassical scatterers. It seems unclear whether it is directly generalizable to generic chaotic systems where no disorder average is involved.[10] In recent work Whitney, Lerner, and Smith [207] have discussed this point. They have analyzed in detail the contributions to the two-level correlator of trajectory pairs of the type shown in Fig. 3.12. As they point out, they could neither correctly reproduce Hikami boxes [208], regions in phase space where quantum processes are important, nor obtain the corrections to the diagonal approximation predicted by random-matrix theory.

To conclude, a complete semiclassical theory for quantum transport in ballistic systems is still lacking.

---

[10] The approach of Aleiner and Larkin has been applied to weak localization in chaotic cavities in [209].

4. Orbital Magnetism

4.1 Historical Backround and Overview

Mesoscopic quantum devices have usually been studied experimentally by connecting them to metal leads and measuring conductance properties. When operating under mesoscopic conditions at sufficiently low temperature their transport behavior may be modified by quantum effects as discussed in Chap. 3.

In the early nineties the first experiments on the magnetic response of small *isolated* ring structures were carried out, which substantially differ from transport measurements and allow for an alternative probe of density-of-states properties. The observation of persistent currents in normal metal rings [210,211], i.e. the orbital magnetic moment in multiply connected geometries, opened up a whole new branch of mesoscopic physics, including *mesoscopic thermodynamics*.

The existence of normal persistent currents in rings was already proposed in the pioneering work of Büttiker, Imry, and Landauer in 1983 [212], which demonstrated that in the presence of a magnetic flux the ground state of a one-dimensional ring exhibits a current flow. The effect is based on a coherent extension of the corresponding wavefunctions around the ring, even if the electron motion is dominated by elastic impurity scatterers [5, 213]. This was the regime of the original measurement in an array of 10^5 copper rings [210]. The use of an ensemble, originally motivated by experimental reasons, brought up important issues about the differences between the canonical and grand canonical ensembles in the mesoscopic regime [214–216], which we shall review in this chapter. The second early experiment mentioned above achieved the observation of persistent currents in single disordered rings [211] and was followed recently by a refined measurement [217].

These experiments and their surprising results have triggered considerable theoretical activities during the last few years, especially because of a serious disagreement between theory and experiment. This disagreement still persists and remains to be understood in spite of important developments in the theoretical understanding.[1] We shall briefly review this issue in Sect. 6.3.2 when we discuss interaction effects in *diffusive* quantum systems and their

---

[1] For recent reviews see [8, 218, 219].

role with regard to the persistent current. In this chapter we shall mainly focus on magnetism in the *ballistic* regime within a picture of noninteracting particles. (The role of electron–electron interactions in ballistic systems will be addressed in Sect. 6.4.)

The semiclassical studies of orbital magnetism [45, 69–71, 221–223], on which this chapter relies, were partly motivated by two later experiments on an ensemble of square quantum dots [57] as well as individual ring microstructures [220] defined on high-mobility semiconductor samples operating in the *ballistic* regime. There impurity scattering is suppressed, which suggests the use of clean models, implying that the electron motion is governed by the confinement potential and geometrical effects play a major role. The experiment on the array of squares showed an unexpectedly large paramagnetic response at zero field (see Fig. 1.5 and discussion in the introduction), which will be addressed in this chapter, too.

In related experiments on individual transition metal and rare-earth element quantum dots, shell effects in the magnetic moment have been observed [224].

The study of orbital magnetism in extended systems as well as the investigation of confinement effects has a long history and goes back to the 1930s with the pioneering work of Landau [112, 225]. He demonstrated the existence of a small diamagnetic response in an electron gas at weak fields H and low temperatures T such that $k_B T$ exceeds the typical spacing $\hbar \omega_{\rm cyc}$. Landau's work was only slowly accepted, for the following reasons [45]: first, it gives a purely quantum result that can be expressed as a thermodynamic relationship without an explicit $\hbar$ dependence. In contrast to this, the Bohr–van Leeuwen theorem [226] establishes the absence of magnetism for a system of classical particles. For finite systems the boundary currents can be shown to exactly cancel in a subtle way the diamagnetic contribution from cyclotron orbits of the interior. Second, boundary effects, which were essential in obtaining the correct classical behavior, did not enter into Landau's derivation. Third, Landau diamagnetism for standard metals yields a small effect (one-third of the Pauli spin paramagnetism), making its experimental observation rather difficult.

The restriction of the electron gas to two dimensions does not open up any new conceptual difficulty [227, 228], but the confinement of the electron system to a finite volume introduces the typical level spacing Δ as a new energy scale into the problem. This leads to a modification of the Landau susceptibility. The latter point has therefore been the object of a long sequence of conflicting studies (reviewed in [229, 230]). The theoretical investigation of finite-size corrections was motivated by experiments on small metal clusters and dealt with various model systems: thin plates [231], thin cylinders [232], confinement by quadratic potentials [233, 234], circular boxes [235], and rectangular boxes [236, 237].

Similarly to the transport properties discussed in the previous chapter, finite-size effects and corrections to bulk magnetism obviously depend on the relation between the typical size a of the system and other relevant length scales [38]: the thermal length L_T, the elastic mean free path l, and the phase coherence length ℓ_ϕ. Most of the above-mentioned studies neglect scattering mechanisms other than that by the boundaries and deal with the macroscopic high-temperature case $L_T \ll a$. In this limit the response is dominated by its smooth component, for which only tiny corrections to the the diamagnetic bulk susceptibility are found [238–240]. Oppositely to the macroscopic limit, there have been studies in the quantum limit $k_B T < \Delta$ [241, 242]. In this regime the magnetic susceptibility is dominated by irregular fluctuations that complicate its unequivocal determination. The purpose of the present chapter is to review studies of size corrections in the mesoscopic regime, $L_T/a > 1 > \beta\Delta$, intermediate between the two previous limits.

A central conclusion of the work [38, 45, 71, 222, 223, 243] on orbital magnetism in the ballistic regime, reviewed here, is that finite-size corrections to the magnetic susceptibility in the *ballistic* regime can be *orders of magnitude* larger than the bulk values. In order to illustrate this effect imagine a mesoscopic square quantum well of size a connected to an electron reservoir with chemical potential μ. A numerical diagonalization of the corresponding Hamiltonian in the presence of a magnetic field yields a diagram of the energy levels as a function of the magnetic flux φ as shown in Fig. 4.1. In between the two separable limiting cases $\varphi = 0$ and $\varphi \longrightarrow \infty$ the spectrum exhibits a complex structure, typical of a nonintegrable system whose classical dynamics is at least partly chaotic. Figure 4.2 displays the corresponding numerically obtained magnetic susceptibility (solid curve).

In the high-field region $(2r_{\mathrm{cyc}} < a)$ characteristic de Haas–van Alphen oscillations are obtained, although not with the amplitude expected from calculations for the bulk (see (4.14)). For lower fields a striking discrepancy is observed between the numerical results, showing pronounced quantum oscillations, and the constant bulk Landau diamagnetism (of order 1 in Fig. 4.2). Thus, the confinement strongly alters the orbital response of an electron gas. As will be shown, the whole curve is well reproduced by a finite-temperature semiclassical theory (dashed line) that takes into account only a few fundamental periodic orbits.

The problem of orbital magnetism from a quantum chaos point of view was first addressed by Nakamura and Thomas [246] in their numerical study of the differences in the magnetic response of circular and elliptic billiards at zero temperature. Since then the question how the character (integrable or chaotic) of the classical dynamics affects magnetic quantities has become a frequently addressed issue. The relevant semiclassical literature, mainly assuming the dynamics of a clean geometry [45, 70, 71, 221–223, 243, 245], will be discussed in the following sections.

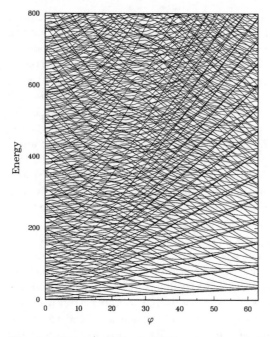

Fig. 4.1. Diagram of the first 200 energy levels of one symmetry class (see Appendix A.2) of a clean square billiard in a uniform magnetic field H as a function of the normalized flux $\varphi = Ha^2/\phi_0$ with $\phi_0 = hc/e$. The energies are scaled such that the zero field limit gives $E = n_x^2 + n_y^2$. (From [45], by permission)

The purpose of this chapter is to review a semiclassical theory of the orbital magnetic properties of noninteracting, spinless electrons in the ballistic regime. We restrict ourselves to the clean limit, where the different behaviors of the magnetic response arise as a geometrical effect (the shape of the microstructure). In Chaps. 5 and 6 we refine the idealized model of clean systems by including disorder and interaction effects.

This chapter, which follows essentially the lines of a related review [45], is organized as follows. We begin with a brief review of bulk magnetic properties showing that Landau diamagnetism is also present in a confined geometry at arbitrary fields. We then present the appropriate thermodynamic formalism to be used in the mesoscopic regime. In Sect. 4.3 we address the magnetic response (susceptibility and persistent current) in *chaotic* systems, deriving a universal semiclassical line shape for the averaged magnetization. In Sect. 4.4 we compare these results with results for *generic integrable* structures, systems whose integrability is broken by the effect of an applied magnetic field. In Sect. 4.5 we present, as a related example, calculations of the magnetic susceptibility for the experimentally relevant case of the square quantum well. The magnetic response of systems such as ring billiards that remain *integrable at arbitrary magnetic field* is addressed in Sect. 4.6.

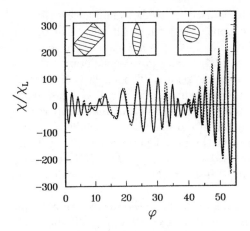

Fig. 4.2. Susceptibility oscillations of a square as a function of the magnetic flux. *Solid curve*: numerical quantum calculations from the energy-level diagram in Fig. 4.1 at finite temperature at an energy corresponding to $\sim$1100 occupied independent-particle states. The susceptibility exhibits pronounced oscillations which are strongly enhanced with respect to the bulk value χ_{L}. They are accurately reproduced by analytical semiclassical expressions (*dashed line*) based on families of quantized flux-enclosing electron orbits, which are shown in the *upper insets* for the different magnetic-field regimes. (From [45], by permission)

4.2 Basic Concepts

We first present the basic thermodynamic relations and definitions and briefly review some results for the bulk electron gas before introducing specific concepts of mesoscopic magnetism.

4.2.1 Definitions

Let us consider a noninteracting electron gas in a volume A (area in two dimensions) at temperature T subject to a static magnetic field H. The magnetic moment of the system in statistical equilibrium is given by the thermodynamic relation

$$\mathcal{M} = -\left(\frac{\partial \Omega}{\partial H}\right)_{T,\mu}, \tag{4.1}$$

where $\Omega(T, \mu, H)$ is the thermodynamic potential (2.38) and μ the chemical potential of the electron gas. The differential magnetic susceptibility is defined by

$$\chi^{\mathrm{GC}} = \frac{1}{A}\left(\frac{\partial \mathcal{M}}{\partial H}\right)_{T,\mu} = -\frac{1}{A}\left(\frac{\partial^2 \Omega}{\partial H^2}\right)_{T,\mu}. \tag{4.2}$$

A notation with the superscript GC is used in order to emphasize the fact that we are working in the grand canonical ensemble. The choice of the ensemble in the macroscopic limit of N and $A \to \infty$ is a matter of convenience. However, as is well known by now [214–216], the equivalence between the ensembles may break down in the mesoscopic regime. This effect is one prominent characteristic of mesoscopic thermodynamics and will be thoroughly discussed in this chapter. For the following discussion of bulk properties it is, however, appropriate to work in the grand canonical ensemble (at fixed μ) and make use of the simple form of the grand potential (2.38) in terms of the single-particle density of states $d(E) = g_s \sum_\lambda \delta(E - E_\lambda)$. The magnetic susceptibility is directly extracted from the knowledge of the density of states.

We shall neglect the Zeeman-splitting term due to the electron spin. It can, however, be incorporated easily when spin–orbit coupling is negligible [247]. Furthermore, orbital magnetism typically dominates over the spin paramagnetic susceptibility; this is usually the case in doped semiconductors.

4.2.2 Bulk Properties: de Haas–van Alphen Effect and Landau Diamagnetism

The case of a free electron gas is particularly simple since the electron eigenstates are Landau states with energies

$$E_k = \hbar\omega_{\text{cyc}} \, (k + 1/2) \qquad\qquad k = 0, 1, 2, \ldots \qquad (4.3)$$

and degeneracies $g_s\phi/\phi_0$, where $\phi = HA$ is the flux through an area A, and $\phi_0 = hc/e$ is the elemental flux quantum.

The magnetic response of a bulk electron gas is characterized by two main features: an overall small diamagnetic response and the well-known de Haas–van Alphen oscillations for magnetic fields H such that the energy scale of the corresponding cyclotron frequency $\omega_{\text{cyc}} = eH/mc$ is larger than the thermal smearing. The computation of the diamagnetism of a free-electron system dates back to Landau's work in 1930 [225]. The derivation, being based on the quantization condition (4.3), can be found for the three-dimensional case in standard textbooks [112, 227]. The two-dimensional case [228–230], where the magnetic field is perpendicular to the plane, follows the same lines and will be sketched here. To this end we employ the Poisson summation

$$\sum_{k=-\infty}^{\infty} g(k) = \sum_{j=-\infty}^{\infty} \chi(j) \, . \qquad (4.4)$$

Here, $\chi(\sigma)$ is the Fourier transform of $g(s)$: $\chi(\sigma) = \int ds \, g(s) \exp(-2\pi i s\sigma)$.

The density of states related to the quantization condition (4.3) can be rewritten by means of (4.4) as

$$d(E, H) = g_s \frac{mA}{2\pi\hbar^2} \left[1 + 2 \sum_{j=1}^{\infty} (-1)^j \cos\left(j \frac{2\pi E}{\hbar\omega_{\text{cyc}}} \right) \right] \, . \qquad (4.5)$$

This decomposition is frequently interpreted as coming from the Weyl term and the strongly field- and energy-dependent contribution of cyclotron orbits. However, at the bottom of the spectra, from which the Landau diamagnetic component originates, this distinction is essentially meaningless.

Landau Susceptibility

For a degenerate electron gas in a weak field such that $\hbar\omega_{\text{cyc}} \ll k_{\text{B}}T \ll \mu$, the energy integral (2.38) gives

$$\Omega(\mu) \simeq \bar{\Omega}(\mu) = -g_{\text{s}}\frac{mA}{2\pi\hbar^2}\frac{\mu^2}{2} + g_{\text{s}}\frac{e^2}{24\pi mc^2}\frac{AH^2}{2} , \qquad (4.6)$$

where $\bar{\Omega}$ denotes the smooth part of the thermodynamic potential (2.38). Upon taking magnetic-field derivatives we thus obtain the two-dimensional diamagnetic Landau susceptibility

$$-\chi_{\text{L}} = -\frac{g_{\text{s}}e^2}{24\pi mc^2} . \qquad (4.7)$$

We note that the diamagnetic response stems from the integral of the rapidly oscillating term of the density of states (4.5) and survives even at high temperatures.

The above derivation explicitly employs the structure of the spectral density in terms of bulk Landau states. In the following we consider the problem from a more general point of view without assuming a specific form of the density of states. This allows us to treat also the magnetic response of confined systems at arbitrary (even weak) magnetic fields. Our semiclassical derivation relies neither, on the quantum side, on the existence of Landau levels, nor, on the classical side, on boundary trajectories or the presence of circular cyclotron orbits fitting into the confinement potential.

In Sect. 2.3 we showed that the various quantum mechanical (i.e. $d(E)$, $n(E)$, $\omega(E)$) and thermodynamic (i.e. $D(\mu)$, $N(\mu)$, $\Omega(\mu)$) spectral properties of a mesoscopic system can be decomposed into smooth and fluctuating parts. In the semiclassical limit, each of these quantities allows an asymptotic expansion in powers of $\hbar$. For most purposes it is sufficient to consider only leading-order terms, and higher-order corrections must be added only if the former vanish for some reason. This is the case for the smooth part $\bar{\Omega}(\mu)$ of the grand potential. It is the dominant term at any temperature, but it is magnetic-field-independent to leading order in $\hbar$ as can be seen in (4.6) for the bulk example. This is the reason for the absence of orbital magnetism in classical mechanics. The computation of field-dependent higher-order $\hbar$ corrections is presented in [45]. It is based on the notion of the Wigner transform (2.20) [85,250] of the Hamiltonian, which is appropriate to studying the smoothed spectral quantities.[2]

---

[2] Analogous results can be found in [238, 251]. The Wigner distribution function was also previously used by Kubo [252] in the study of bulk Landau diamagnetism.

To observe a field dependence, one must consider the leading-order field-dependent correction to the Wigner transform, $\bar{\Omega}_W(\mu)$, of the grand potential. It reads [45]

$$\bar{\Omega}_1(\mu) = \frac{\mu_B^2 H^2}{6} \, \bar{D}_W(\mu) \, . \tag{4.8}$$

In the grand canonical ensemble, the above equation readily gives the susceptibility

$$\bar{\chi}^{GC} = -\frac{\mu_B^2}{3A} \, \bar{D}_W \, , \tag{4.9}$$

where $\bar{D}_W$ is the Weyl part of the thermodynamic density of states (2.40a). Using the relation $\bar{D}_W = d\bar{N}_W/d\mu$, one recognizes the familiar result of Landau [225]. For systems without a potential (bulk or billiard systems) it gives, in the degenerate case ($\mu \gg k_B T$) in two or three dimensions,

$$\bar{\chi}_{2d}^{GC} = -\frac{g_s e^2}{24\pi mc^2} \, , \qquad \bar{\chi}_{3d}^{GC} = -\frac{g_s e^2 k_F}{24\pi^2 mc^2} \, . \tag{4.10}$$

The susceptibility reads, in the nondegenerate limit,

$$\bar{\chi}^{GC} = -\frac{\mu_B^2}{3A} \, \frac{N}{k_B T} \, , \tag{4.11}$$

where N is the mean particle number. The Landau contribution is dominant at high temperatures: it is temperature-independent in the degenerate regime and shows a $1/T$ decay in the nondegenerate limit, while additional oscillatory contributions, to be studied in the remainder of this section, are then exponentially damped by temperature.

The Landau diamagnetism is usually derived for free electrons or for a quadratic confining potential [112, 227]. The derivations in [45, 238] provide generalizations to any confining potential (including systems smaller than the cyclotron radius) since the confining potential $V(\boldsymbol{r})$ does not enter into the leading-order field-dependent terms of $\overline{\Omega}(\mu)$. This shows that the Landau susceptibility is a property of infinite systems as well as mesoscopic devices. The Landau diamagnetic response is the same in the canonical case. The contribution from electron interactions can be neglected (see Chap. 6).

De Haas–van Alphen Oscillations

So far we have discussed the magnetic response at high temperature. Consider now the oscillating contributions to the density of states, (4.5), of infinite systems at low temperatures such that $k_B T \ll \hbar\omega_{cyc}$. The energy integral (2.38) over the rapidly oscillating component is not negligible. Using a relation similar to (2.47) this integral gives

$$\Omega^{osc} = \frac{g_s mA}{\pi\hbar^2} \sum_{j=1}^{\infty} (-1)^j \left(\frac{\hbar\omega_{cyc}}{2\pi j}\right)^2 \cos\left(\frac{2\pi j\mu}{\hbar\omega_{cyc}}\right) R\left(\frac{2\pi j r_{cyc}}{L_T}\right) \, . \tag{4.12}$$

The temperature-dependent damping factor (2.49) containing the thermal cutoff length L_T (see 2.50) reads in this case

$$R\left(\frac{j2\pi r_{\text{cyc}}}{L_T}\right) = \frac{2\pi^2 j k_\text{B} T/\hbar\omega_{\text{cyc}}}{\sinh\left(2\pi^2 j k_\text{B} T/\hbar\omega_{\text{cyc}}\right)} . \tag{4.13}$$

Combining both Landau and de Haas–van Alphen contributions, the orbital magnetic susceptibility for the clean bulk reads [45]

$$\frac{\chi^{\text{GC}}}{\chi_\text{L}} = -1 - 24\left(\frac{\mu}{\hbar\omega_{\text{cyc}}}\right)^2 \sum_{j=1}^{\infty}(-1)^j \cos\left(j\frac{2\pi\mu}{\hbar\omega_{\text{cyc}}}\right) R\left(j\frac{2\pi r_{\text{cyc}}}{L_T}\right). \tag{4.14}$$

The second term exhibits the characteristic oscillations with period $1/H$ and is exponentially damped with temperature.

The effect of weak random disorder on the de Haas–van Alphen oscillations will be computed in Sect. 5.3. There we also compare the result with a susceptibility measurement of a 2DEG by Eisenstein et al. [248]. We note that for high fields one cannot in principle separate the orbital and spin effects.

4.2.3 Thermodynamics in the Mesoscopic Regime

While going from the bulk two-dimensional case (macroscopic regime) to the constrained case (ballistic mesoscopic regime) two important changes take place: the confining energy appears as a relevant scale and (4.3) no longer provides the quantization condition; furthermore, since we are not in the thermodynamic limit of N and $A \to \infty$, the constraint of a constant number of electrons in an isolated microstructure is no longer equivalent to having a fixed chemical potential. Referring to the second point, we shall present the thermodynamic framework and introduce semiclassical concepts which deal appropriately with this situation.

In contrast to an infinite system or the situation (in transport) where a mesoscopic device is connected to a reservoir of particles with chemical potential μ, the number N of particles inside an *isolated* microstructure is fixed, although it may be rather large (of order 10^5) in actual quantum dots. It is essential in some cases, namely when considering the average susceptibility of an ensemble of microstructures, to take into account explicitly this conservation of N and to work within the canonical ensemble. For such systems with a fixed number of particles, the relevant thermodynamic function is the free energy F, the Legendre transform of the grand potential Ω:[3]

$$F(T, H, N) = \mu N + \Omega(T, H, \mu) . \tag{4.15}$$

Then, the magnetization $\mathcal{M}$ of an isolated system of N electrons, which for ring geometries is usually expressed in terms of the persistent current I, is given by the thermodynamic relation

---

[3] For a justification of this relation in the mesoscopic context, see [45].

$$I = \frac{c}{A} \, \mathcal{M} = -c \, \left(\frac{\partial F}{\partial \phi} \right)_{T,N} \tag{4.16}$$

and the canonical magnetic susceptibility is defined as

$$\chi = -\frac{1}{A} \left(\frac{\partial^2 F}{\partial H^2} \right)_{T,N} . \tag{4.17}$$

Throughout this book we shall denote the energy- or size-*averaged* persistent current and susceptibility of an ensemble of mesoscopic structures by $\overline{I}$ and $\overline{\chi}$, respectively. The *typical* current and susceptibility are defined as

$$I^{(t)} = \sqrt{\overline{I^2}} \quad ; \quad \chi^{(t)} = \sqrt{\overline{\chi^2}} . \tag{4.18}$$

These quantities apply to the case of repeated measurements on a given microstructure, when variations in k_F are obtained by some kind of perturbation, and serve as a measure of the variance in the magnetic response. For additional disorder averages (discussed in Chap. 5) we use the notation $\langle \cdots \rangle$.

In Sect. 2.3 we derived semiclassical expressions for the smoothed (single-particle) density of states $D(\mu)$, the staircase function $N(\mu)$, and the grand potential $\Omega(\mu)$, which provide a natural starting point for calculating the mesoscopic orbital magnetic response. While the grand canonical quantities follow directly from derivatives of $\Omega(\mu)$, the semiclassical calculation of the free energy is not obvious. To achieve this we follow [45], which is based on Imry's derivation for persistent currents in ensembles of disordered rings [215]. The only important difference is that we shall take averages over the size and the Fermi energy of ballistic structures instead of averages over impurity realizations.

As mentioned above, the defining equation (4.17) of the canonical susceptibility χ is equivalent to χ^{GC} (see (4.2)) up to corrections of order $1/N$ (i.e. $\hbar$). Therefore, in the mesoscopic regime of small structures with large but finite N we have to consider such corrections if we want to take advantage of the computational simplicity of the grand canonical ensemble (GCE). The difference between the two definitions is particularly important when the GCE result is zero, apart from the Landau diamagnetic contribution, as is the case for the ensemble average of χ^{GC} as shown below. The computation of the correction terms can be achieved from the relationship (4.15) between the thermodynamic functions $F(N)$ and $\Omega(\mu)$ and the relation $N(\mu) = N$. In the case of finite systems the previous implicit relation is difficult to invert. However, when N is large we can use the decomposition of $N(\mu)$ into a smooth part $\overline{N}(\mu)$ and a small fluctuating component $N^{osc}(\mu)$, (2.54b). This allows for a perturbative treatment of the previous implicit relation. The contribution of a given orbit to d^{osc} is always of lower order in $\hbar$ than $\overline{d}$, as can be checked by inspection of the various semiclassical trace formulas. However, since there are infinitely many such contributions, d^{osc} and $\overline{d}$ are of the same order when adding them up. Indeed, this must be the case since the quantum mechanical $d(E)$ is a sum of δ peaks. Hence one cannot simply use

$d^{osc}/\bar{d}$ as a small expansion parameter. However, finite temperature provides an exponential cutoff in the length of the trajectories contributing to D^{osc}, so that only a finite number of them must be taken into account. Therefore, D^{osc} is of lower order in $\hbar$ than $\bar{D}$, and in the semiclassical regime it is possible to expand the free energy F with respect to the small parameter $D^{osc}/\bar{D}$. The use of a temperature-smoothed density of states therefore justifies this approach.

In order to perform the perturbative expansion sketched above we define a mean chemical potential $\bar{\mu}$ by introducing the condition

$$N = N(\mu) = \bar{N}(\bar{\mu}) \ . \tag{4.19}$$

Figure 4.3 illustrates this relation for the two-dimensional (potential-free) case where $\bar{D}$ is constant. The expansion of the above relation to first order in $D^{osc}/\bar{D}$ gives, employing the fact that $dN/d\mu = D$,

$$\Delta\mu \equiv \mu - \bar{\mu} \simeq - \frac{1}{\bar{D}(\bar{\mu})} \ N^{osc}(\bar{\mu}) \ . \tag{4.20}$$

The interpretation of $\Delta\mu$ is seen in Fig. 4.3: the shaded area represents the number of electrons in the system and it is equal to the product $\bar{D}\bar{\mu}$.

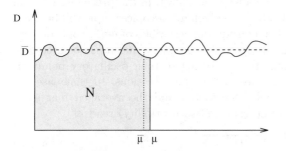

Fig. 4.3. Sketch of the decomposition of the thermodynamic density of states $D(\mu)$ (*solid line*) into a smooth part $\bar{D}$ (*dashed line*) and an oscillating component. The total number of electrons N is indicated by the *shaded area*, and equal to the product of $\bar{D}$ and $\bar{\mu}$. (From [45], by permission)

Expanding the relationship (4.15) to second order in $\Delta\mu$ gives

$$F(N) \simeq (\bar{\mu} + \Delta\mu)N + \Omega(\bar{\mu}) - N(\bar{\mu})\Delta\mu - D(\bar{\mu}) \frac{\Delta\mu^2}{2} \ . \tag{4.21}$$

Upon using the decomposition of $\Omega(\bar{\mu})$ and $N(\bar{\mu})$ into mean and oscillating parts, (2.54b) and (2.54c), and eliminating $\Delta\mu$ (4.20) in the second-order term, one obtains the expansion of the free energy to second order in $D^{osc}/\bar{D}$ [45, 215, 216]

$$F(N) \simeq F^0 + \Delta F^{(1)} + \Delta F^{(2)} \ , \tag{4.22}$$

with

$$F^0 = \bar{\mu} \boldsymbol{N} + \bar{\Omega}(\bar{\mu}) \, , \tag{4.23a}$$

$$\Delta F^{(1)} = \Omega^{\mathrm{osc}}(\bar{\mu}) \, , \tag{4.23b}$$

$$\Delta F^{(2)} = \frac{1}{2\bar{D}(\bar{\mu})} \, [N^{\mathrm{osc}}(\bar{\mu})]^2 \, . \tag{4.23c}$$

The terms $\Delta F^{(1)}$ and a $\Delta F^{(2)}$ can be expressed through the oscillating part of the density of states by means of (2.54b) and (2.54c). The first two terms, F^0 and $\Delta F^{(1)}$, yield the magnetic response calculated in the GCE at the effective chemical potential $\bar{\mu}$. $\Delta F^{(2)}$ represents the leading-order difference between the grand canonical and canonical results. It is of grand canonical form since it is expressed in terms of an integral of the density of states (2.37) for a fixed chemical potential $\bar{\mu}$. It is convenient to use the expansion (4.22) in the calculation of the magnetic susceptibility of a system with a fixed number of particles, because the leading $\hbar$ contribution to $\boldsymbol{N} = \bar{N}(\bar{\mu})$ has no magnetic-field dependence, independent of the precise system under consideration. At this level of approximation, keeping $\boldsymbol{N}$ constant in (4.17) when taking the magnetic-field derivative amounts to keeping $\bar{\mu}$ fixed.

The term $F^{(0)}$ contains only the small diamagnetic Landau susceptibility arising from a higher-order $\hbar$ expansion, as shown in the preceding section. Hence, the weak-field susceptibility of a given mesoscopic sample will be dominated by $\Delta F^{(1)}$. However, when considering ensembles of mesoscopic devices with slightly different sizes or electron fillings, $\Delta F^{(1)}$ and its associated susceptibility contribution average to zero owing to their oscillatory behavior. This will be explicitly demonstrated in the next sections. In that case the next-order term $\Delta F^{(2)}$ has to be considered and the averaged magnetic-response quantities (persistent current and susceptibility) read

$$\bar{I}(\bar{\mu}, \phi) \simeq -\frac{c}{2\bar{D}(\bar{\mu})} \frac{\partial}{\partial \phi} \overline{[N^{\mathrm{osc}}(\bar{\mu}, \phi)]^2} \, , \tag{4.24a}$$

$$\bar{\chi}(\bar{\mu}, H) \simeq -\frac{1}{2A\bar{D}(\bar{\mu})} \frac{\partial^2}{\partial H^2} \overline{[N^{\mathrm{osc}}(\bar{\mu}, H)]^2} \, . \tag{4.24b}$$

In the following we compute semiclassical expressions for $\Delta F^{(1)}$ and $\Delta F^{(2)}$ for the different cases of systems with chaotic and integrable classical dynamics.

4.3 Chaotic Systems

We begin our analysis of the magnetic properties of ballistic quantum systems with those possessing classically chaotic dynamics. Each of these systems exhibits certain specific properties depending on size and geometry, which enter, for instance, into the (absolute) magnitude of its magnetic response. However, the fact that their classical dynamics is supposed to be completely chaotic implies generic features which characterize uniquely and rather generally a broad class of mesoscopic structures. Here we outline a semiclassical theory which, on the one hand, aims at the generic character in terms of universal magnetic correlation functions. These are determined by evaluating semiclassical trace formulas using statistical assumptions for the periodic orbits. On the other hand, this approach allows one to incorporate specific knowledge of classical phase space properties to obtain even (nonnormalized) absolute results.

According to (4.23c) and its magnetic-field derivatives, the averaged magnetic response probes directly the flux sensitivity of quantum fluctuations in the particle number $(N^{\mathrm{osc}})^2$ of a quantum dot at low temperatures. These fluctuations are closely related to two-point correlations in the density of states (2.28). The magnetic response functions for chaotic systems therefore represent a prominent and physically relevant class of universal parametric correlation functions, to be discussed below.

The outline of this section is as follows: after a brief review of the present semiclassical approaches to orbital magnetism of ballistic chaotic systems we focus on persistent-current and susceptibility correlation functions. The resulting semiclassical predictions for ensembles of chaotic structures will be compared with corresponding exact quantum calculations performed for a cylinder geometry with classical chaotic dynamics [70]. The system considered allows one to access a regime of large Fermi momenta k_{F} and enables one to address issues of "self-averaging" [254] and the relation between individual chaotic systems and ensembles of (disordered) systems [46]. In the remainder of this section we generalize the results to the case of singly connected ballistic quantum dots [45], compare these with the diffusive case, and discuss possible experimental implications.

4.3.1 Semiclassical Approaches

The concept of employing magnetic properties in order to characterize clean mesoscopic systems with chaotic dynamics has been addressed in several publications during the last few years: Nakamura and Thomas [246] were the first to investigate orbital magnetism from a quantum chaos point of view. They studied numerically the differences in the magnetic response of circular and elliptic billiards. The former remain integrable, while the latter develop chaotic behavior at finite magnetic fields. These authors found a reduction compared to the bulk susceptibility and strong fluctuations with varying magnetic field and observed that both effects were stronger for the elliptic billiard. The difficulty of these studies at zero temperature consists in the existence of strong fluctuations arising from exact crossings or quasi-crossings of energy levels (depending parametrically on the magnetic field) where the susceptibility diverges. These fluctuations mask the generic differences between integrable and chaotic systems related to a different $\hbar$ dependence, to be discussed in Sect. 4.4.2. In this quantum limit of very low temperatures, which no longer provide a cutoff in orbit lengths, one is moreover faced with convergence problems of the trace formulas as mentioned in Sect. 2.1.4. Agam [222] therefore introduced Berry–Keating resummation techniques [255] and calculated the susceptibility of individual systems on the basis of rearranged semiclassical expressions for the density of states.

Berry and Keating [245], using semiclassical asymptotics, computed the flux- and energy-averaged autocorrelation function

$$C(\phi) = \bar{d}^2 \int_0^1 d\phi_0 \left\langle \frac{dE_n}{d\phi}(\phi_0 + \phi) \frac{dE_n}{d\phi}(\phi_0) \right\rangle \tag{4.25}$$

for ring geometries threaded by a flux. This function was originally introduced by Szafer and Altshuler [257]. It correlates level velocities at flux values differing by ϕ. By evaluating related semiclassical trace formulas and employing certain statistical properties of the classical orbits and their winding numbers, they found the uniform approximation

$$C(\phi) \approx -\frac{\sin^2(\pi\phi) - 1/n^{*2}}{[\sin^2(\pi\phi) + 1/n^{*2}]^2}, \tag{4.26}$$

where n^* is the RMS winding number of orbits with periods of the Heisenberg time. For large ϕ (4.26) reaches the asympthotic form $C(\phi) \approx -1/\pi^2\phi^2$ derived by Szafer and Altshuler [257].

The function $C(\phi)$, correlating levels at *different* fluxes, differs, however, from magnetic response functions which are based on flux derivatives of the thermodynamic relations (4.23b) and (4.23c). A semiclassical analysis of the latter was performed by several authors: treating periodic orbits statistically, Serota [256] derived expressions for the averaged and typical persistent currents at zero temperature. Von Oppen and Riedel [243], calculating the persistent current of a single Sinai billiard, pointed out the different parametric dependences of its magnitude on k_F for systems with mixed and purely chaotic phase space.

Related concepts were used to compute the magnetic response at finite temperature for Aharonov–Bohm geometries enclosing a flux line [70] as well as chaotic systems in a uniform field [45,222]. In the following, these analytical computations will be presented and compared with accurate numerical calculations. In particular, we discuss the role of temperature smoothing in the flux correlation functions and suggest it as a natural and physically relevant parameter for studying (experimentally) the signatures of different semiclassical timescales in smoothed spectral correlation functions. Contrary to the analysis of Berry and Keating, who estimated off-diagonal contributions to $C(\phi)$ to be negligible, we observe the onset of the breakdown of the semiclassical diagonal approximation for a smoothing smaller than the mean level spacing Δ.

Let us consider systems where all the periodic orbits are sufficiently isolated that the trace of the semiclassical Green function (2.18) can be evaluated within the stationary-phase approximation. In this case the Gutzwiller trace formula (2.27) provides the appropriate starting point for calculating the oscillating part of the density of states, $d^{osc}(E, H)$. This enters into the expressions for N^{osc} and Ω^{osc}, (2.54b) and (2.54c), which are required for the calculation of the averaged (4.24a) and typical (4.18) magnetic response.

All the classical quantities in the oscillating contributions $d_{po}(E, H)$ given by (2.27) associated with periodic orbits generally depend on energy and magnetic field. However, owing to the structural stability of chaotic systems, small changes in H will generally not change the phase space properties and the orbits involved; the system will remain hyperbolic. The zero-field behavior is therefore not substantially different from that at finite fields as far as the stability of the dynamics is concerned. Therefore, we do not need to restrict ourselves to weak fields. In this respect chaotic geometries have the same conceptual simplicity as systems which remain integrable at arbitrary fields to be studied in Sect. 4.6.

The field dependence of each contribution $d_{po}(E, H)$ to the oscillating part of the density of states appears essentially because of the modification of the classical actions since they enter, multiplied by the large factor $1/\hbar$, as rapidly varying phases. To leading order in $\hbar$, the other classical entries such as periods and stability factors can thus be regarded as effectively field-independent. According to classical perturbation theory and (2.15), we represent changes in the action due to the change δH of the field (to first order) by

$$\frac{\partial S_{po}(H)}{\partial H} = \frac{e}{c} \mathcal{A}_{po}(H) , \qquad (4.27)$$

where $\mathcal{A}_{po}(H)$ is the directed area enclosed by the periodic orbit at the field considered. Within this approximation taking the field derivative of terms in periodic-orbit trace formulas (2.27) essentially amounts to a multiplication by factors $\pm(e/c\hbar)\mathcal{A}_{po}$.

In the special case of the magnetic response to a weak field one can employ time reversal invariance and express $d_{po}(E, H)$ in terms of the characteristics of the orbits at zero field: grouping together each (nonself-retracing) orbit with its time-reversed counterpart yields (see also (4.95))

$$d_{po}(E, H) \simeq d_{po}^0 \cos\left(2\pi \frac{H \mathcal{A}_{po}^0}{\phi_0}\right) . \qquad (4.28)$$

Here, d_{po}^0 is the contribution of the *unperturbed* orbit obtained from (2.27) at $H = 0$, and $\mathcal{A}_{po}^0$ is its enclosed area. $\phi_0 = hc/e$ stands for the flux quantum.

For ring geometries the above relations are exact since the magnetic field is considered only to create a flux which affects the phases of quantum objects but not the related classical dynamics. In the following we shall first study semiclassically and quantum mechanically persistent currents in rings and then also present results for susceptibilities of general quantum billiards in uniform fields.

4.3.2 Persistent Currents

For ring topologies it is convenient to measure the flux enclosed by a periodic orbit in terms of its winding number n_{po} counting the net number of revolutions around the ring: $H\mathcal{A}_{\mathrm{po}} = n_{\mathrm{po}}\phi$. The classical action (of a particle in a ring billiard) is then given by

$$S_{\mathrm{po}}(E, \varphi) = \hbar(kL_{\mathrm{po}} + 2\pi\, n_{\mathrm{po}}\, \varphi) \tag{4.29}$$

with $\varphi = \phi/\phi_0$ and L_{po} being the orbit length.

The dominant (grand canonical) contribution to the persistent current of a single ring of radius r is obtained by taking the flux derivative of $\Delta F^{(1)}$ (given by (4.23b) and (2.54c)):

$$\frac{I^{(1)}}{I_0} = -\frac{c}{I_0}\frac{\partial \Delta F^{(1)}}{\partial \phi} \tag{4.30}$$

$$= \frac{g_s}{\pi}\sum_{\mathrm{po}}\frac{n_{\mathrm{po}}\,R(L_{\mathrm{po}}/L_T)}{(j_{\mathrm{po}}L_{\mathrm{po}}/r)\,|\det(\boldsymbol{M}_{\mathrm{po}} - \boldsymbol{I})|^{1/2}}\sin\left(\frac{S_{\mathrm{po}}}{\hbar} - \sigma_{\mathrm{po}}\frac{\pi}{2}\right).$$

The persistent current is measured in units of $I_0 = ev_{\mathrm{F}}/2\pi r$, the current of electrons in a one-dimensional clean ring. j_{po} is the number of repetitions that each periodic orbit includes. $I^{(1)}$ can be paramagnetic or diamagnetic with equal probability. The same holds for integrable rings (see Sect. 4.6.1). The function $R(L_{\mathrm{po}}/L_T)$ (see (2.49)) accounts, as usual, for temperature smoothing, leading to a suppression of contributions of long paths on the scale of L_T.

The energy-averaged response of an ensemble of ring structures is given by means of (4.24a) through the flux derivative of $\Delta F^{(2)} = [N^{\mathrm{osc}}(\bar{\mu})]^2/2\bar{D}(\bar{\mu})$, since the grand canonical contribution $I^{(1)}$, containing phases $k_{\mathrm{F}}L_{\mathrm{po}} \gg 1$ in the semiclassical limit, averages to zero.

Using (2.54b) for $N^{\mathrm{osc}}(\bar{\mu})$, its flux derivative is a double sum over all pairs of orbits:

$$\frac{\partial}{\partial\varphi}(N^{\mathrm{osc}})^2 = \frac{2g_s^2}{\pi\phi_0}\sum_{\mathrm{po}',\mathrm{po}}\frac{n_{\mathrm{po}}\,R(L_{\mathrm{po}}/L_T)R(L_{\mathrm{po}'}/L_T)}{j_{\mathrm{po}}j_{\mathrm{po}'}\,|\det(\boldsymbol{M}_{\mathrm{po}} - \boldsymbol{I})\det(\boldsymbol{M}_{\mathrm{po}'} - \boldsymbol{I})|^{1/2}}$$

$$\times\left[\sin\left(\frac{S_{\mathrm{po}} - S_{\mathrm{po}'}}{\hbar} - (\sigma_{\mathrm{po}} - \sigma_{\mathrm{po}'})\frac{\pi}{2}\right)\right.$$

$$\left. + \sin\left(\frac{S_{\mathrm{po}} + S_{\mathrm{po}'}}{\hbar} - (\sigma_{\mathrm{po}} + \sigma_{\mathrm{po}'})\frac{\pi}{2}\right)\right]. \tag{4.31}$$

Here we neglect, as our major approximation, contributions from off-diagonal pairs of orbits that are not related to each other by time reversal symmetry. We therefore assume that they do not survive ensemble averaging. As discussed in Sect. 2.2, the diagonal approximation is justified as long as timescales larger than the Heisenberg time do not play a role (for example as a result of a temperature cutoff). However, without such a cutoff, near-degeneracies in the actions of long nonidentical orbits may appear owing to the exponential proliferation of the number of periodic orbits in chaotic systems and possible correlations between their actions [109]. Those contributions do not average out, as we shall see below when comparing the results of the theory with numerical calculations for low temperatures.

Among the diagonal contributions which persist upon ensemble averaging, only pairs consisting of an orbit po and its time-reversed partner $T(\text{po})$ contribute to the persistent current: they enclose an effective flux 2φ. One then finds, using (4.31),

$$\frac{\partial}{\partial\varphi}\overline{(N^{\text{osc}})^2} \simeq -\frac{4g_s^2}{\pi}\sum_{\text{po}}\frac{R^2(L_{\text{po}}/L_T)}{j_{\text{po}}^2\,|\det(\boldsymbol{M}_{\text{po}}-\boldsymbol{1})|}\,|n_{\text{po}}|\sin\left(4\pi|n_{\text{po}}|\varphi\right),(4.32)$$

where each element of the sum refers to the pair contribution of two periodic orbits po and $T(\text{po})$.

For temperatures such that the thermal length L_T entering the cutoff function is comparable to the system size, the persistent current is dominated by a few of the shortest periodic orbits. It therefore exhibits system-specific features. However, for lower temperatures (increasing L_T) an exponentially increasing number of trajectories will contribute to $\bar{I}$. This allows for a statistical treatment of the sum in (4.32) and the derivation of a *universal* line shape for the averaged persistent current. For sake of clarity, we refer in the following to the case of billiard-like structures, but the following approach can be generalized to systems with (smooth) potentials. The statistical assumptions enter on two levels.

First, assuming a uniform distribution of the periodic orbits in phase space, the Hannay–Ozorio de Almeida sum rule [27] (see (2.34)) enables one to replace the sum (4.32) by an integral over periods or orbit lengths.

The second ingredient is the distribution of areas enclosed by long trajectories. Since this distribution has a generic form [193, 258] for arbitrary chaotic billiards, we first discuss it in general and then consider the specific case of ring geometries. This result follows from a general argument [193]: with a convenient choice of the origin, the "area" swept by the ray vector between two successive bounces on the billiard boundary follows a distribution with zero mean value and a width σ_N. For a strongly chaotic system successive bounces can be taken as independent events. Thus, the probability $P_N(\Theta)\mathrm{d}\Theta$ for a trajectory to enclose an accumulated algebraic area between Θ and $\Theta+\mathrm{d}\Theta$ after N bounces follows from a random-walk process. It is then given by

$$P_N(\Theta) = \frac{1}{\sqrt{2\pi N\sigma_N}}\exp\left(-\frac{\Theta^2}{2N\sigma_N}\right). \tag{4.33}$$

Owing to the central limit theorem the variance is of order $\sqrt{N}$. Denoting by $\bar{L}$ the average distance between two successive reflections and normalizing with $\sigma_L = \sigma_N/\bar{L}$, one finds the same Gaussian law as (4.33) but with N replaced by L. For systems of ring topology Θ is conveniently expressed in terms of winding numbers ($\Theta = n\pi r^2$). The corresponding distribution reads

$$P_L(n) = \frac{1}{\sqrt{2\pi L\sigma}} \exp\left(-\frac{n^2}{2\sigma L}\right) \tag{4.34}$$

for trajectories of a given length L in a chaotic ring billiard.

For temperatures sufficently low that the contributing orbits are long enough to cover phase space uniformly and fulfill the Gaussian winding-number law, one can replace the sum (4.32) by the following integral expression (neglecting higher repetitions j_{po}):

$$\frac{\partial}{\partial\varphi}\overline{(N^{\mathrm{osc}})^2} \simeq -\frac{4g_s{}^2}{\pi} \int_0^\infty \mathrm{d}L \frac{R^2(L/L_T)}{L} \sum_{n=-\infty}^\infty P_L(n)\, n\, \sin\left(4\pi n\varphi\right). \tag{4.35}$$

In a final step, the Poisson summation formula is applied to the sum over winding numbers. Insertion into (4.24a) yields a semiclassical expression for the averaged persistent current of a chaotic system [70]:

$$\frac{\overline{I}}{I_0} \simeq \frac{g_s}{\pi^2}\frac{\Delta}{k_B T}\, a\sigma \sum_{m=-\infty}^\infty (2\varphi - m) \int_0^\infty \mathrm{d}\xi\, R^2(\xi) \exp\left[-g_m(\varphi,\xi)\right]. \tag{4.36}$$

Here, $\xi = L/L_T$,

$$g_m(\varphi,\xi) = 2\pi^2(2\varphi - m)^2 L_T \sigma\, \xi\,, \tag{4.37}$$

and $a = 2\pi r$ denotes the circumference of the ring.

A corresponding semiclassical expression for the typical current $I^{(t)} = [(I^{(1)})^2]^{1/2}$ can be derived along the same lines from (4.30). The result reads

$$\frac{I^{(t)}}{I_0}a \simeq \frac{g_s}{\sqrt{2\pi}} \left\{\frac{a^2\sigma}{L_T} \sum_{m=-\infty}^\infty \int_0^\infty \mathrm{d}\xi \frac{R^2(\xi)}{\xi^2}\right. \tag{4.38}$$

$$\left.\times \left[[1 - g_m(0,\xi)]\mathrm{e}^{-g_m(0,\xi)} - [1 - g_m(\varphi,\xi)]\mathrm{e}^{-g_m(\varphi,\xi)}\right]\right\}^{1/2}.$$

The range of validity of this expression for $I^{(t)}$ is smaller than that for $\overline{I}$ owing to the ξ^{-2} factor, which strongly suppresses long orbits. It diverges at short lengths, where the statistical assumptions made in the derivation begin to break down. Indeed, comparisons with quantum calculations show that the introduction of a short length scale $L_< \simeq a$, cutting off contributions to the integral for $L < L_<$, is required to obtain reasonable agreement.

In the following we compare the semiclassical predictions, derived above for the averaged current with quantum calculations for the case of a billiard as depicted in Fig. 4.4b closed to form a cylinder of height w and circumference a (Fig. 4.4a). The two semidisks give rise to strongly chaotic classical dynamics. In the actual calculation they are larger than indicated in the figure in order to prevent direct, marginally stable flux-enclosing paths. Fig. 4.4c shows a comparison of numerically obtained classical winding-number distributions (histograms) with the Gaussian distributions (according to (4.34)) for two different trajectory lengths. The variance σ of the Gaussian curves

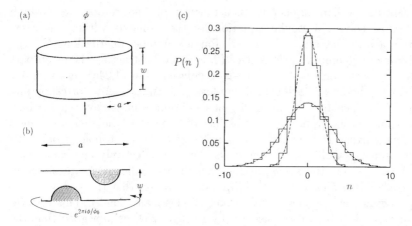

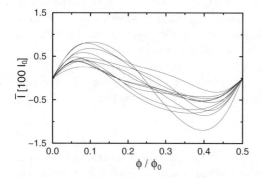

Fig. 4.4. (a) Schematic illustration of a mesoscopic system with the topology of a hollow cylinder threaded by a magnetic flux ϕ in the axial direction. (b) The model system: the two sides of a ballistic billiard with two semidisk scatterers (giving rise to chaotic motion) are attached to each other to form a cylinder. (c) Distribution $P_L(n)$ of the winding numbers n of classical trajectories for the billiard in (b). Classical simulations (*histograms*) are compared with Gaussian distributions (*dashed lines*) obtained from (4.34) for two different trajectory lengths $L/w = 20$ and 90. (From [70], by permission)

was determined from a numerical calculation of the classical diffusion constant D of the system using $\sigma = Dm/\hbar k a^2$ [70, 261]. The excellent agreement justifies the Gaussian winding-number assumption used in (4.34). Although the overall law is universal, the particular value of the diffusion constant D, governing the accumulation of flux, introduces a system-specific scale.

Fig. 4.5. Energy-averaged persistent currents as a function of the magnetic flux $\varphi = \phi/\phi_0$ for chaotic ballistic rings of the same topology (shown in Fig. 4.4) but with slightly differing disk radii. The system is not self-averaging for a energy window corresponding to about 200 levels. (Adapted from [70], by permission)

Quantum mechanical energies as a function of flux are obtained to high accuracy by solving a secular equation for the S-matrix within a scattering approach to quantization [70, 261]. Particle number fluctuations $N^{\mathrm{osc}}(\bar{\mu})$ are obtained by computing $N(\bar{\mu})$ (see (2.37)) and subtracting the mean part $\bar{N}(\bar{\mu})$. The magnetic response is then calculated from (4.24a). It turns out that a pure energy average over a window $[\bar{\mu} - \Delta\mu, \bar{\mu} + \Delta\mu]$ corresponding to up to 200 eigenstates is not sufficient to obtain a unique averaged persistent current: as shown in Fig. 4.5, distinct differences in the *energy-averaged* persistent current $\bar{I}(\varphi)$ appear for samples of the same topology but with slightly differing radii of the semidisks (keeping effectively the same classical dynamics, i.e. the same variance σ). Similar differences are observed when considering $\bar{I}$ as a function of $\bar{\mu}$. This indicates that the system is *not* self-averaging (on an energy scale which is classically small). This obervation supports a point raised recently by Prange [254], who argued that the form factor is not self-averaging. We can therefore infer that, at least for this system, differences occur between the spectral correlation functions of an energy-averaged single chaotic system and a disorder ensemble average. This behavior does not support recent statements on the equivalence between individual chaotic systems and ensembles of (disordered) systems [46].

In order to obtain unique averaged quantum magnetic response functions we have performed an additional average $\langle \ldots \rangle$ over about 30 geometries with slightly differing semidisks. In Fig. 4.6 the results for the averaged quantum mechanical persistent current (full lines) are compared with the semiclassical prediction (dashed lines, (4.36)) over one period of the magnetic flux. The different pairs of curves belong to different temperature smoothings $k_{\mathrm{B}}T/\Delta$ between 0.25 and 4 (the curves with the lowest maximum belong to the highest T). We observe an excellent agreement down to temperatures of about half a level spacing. We stress that we are comparing here *absolute* (not normalized) spectral correlation functions free of any adjustable parameters. The semiclassical theory presented here provides universal persistent-current line shapes. Unlike random-matrix theory, it further allows one to include information on the classical dynamics of the specific system via the diffusion constant σ. In the present case, σ is the sole parameter which determines the absolute heights and widths of the correlation functions.

For the present system (Fig. 4.4) the relation (2.51) between the thermal cutoff length and temperature smoothing can be expressed as

$$\frac{L_T}{a} \simeq \frac{\alpha k_{\mathrm{F}} w}{2\pi^2} \frac{\Delta}{k_{\mathrm{B}}T} , \tag{4.39}$$

where $\alpha \simeq 0.5$ is the ratio of the area of the semidisks to the billiard area aw. The good agreement with the quantum results at rather high temperatures $k_{\mathrm{B}}T/\Delta \simeq 4$ indicates that the statistical assumptions entering into the semiclassical approach hold up to rather short corresponding cutoff lengths $L_T \simeq a$.

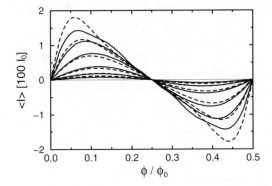

Fig. 4.6. Averaged persistent current for an ensemble of chaotic ballistic rings (see Fig. 4.4) over one period of the magnetic flux $\varphi = \phi/\phi_0$ for different temperatures ($k_B T/\Delta = 0.25, 0.5, 1, 2, 3, 4$ from the *top*, for the *left maximum*) at $k_F w \simeq 90$. Semiclassical results (*dashed lines*) from expression (4.36) are compared with those from exact quantum calculations (*full lines*). (From [70], by permission)

In addtition, the low-temperature limit is of special theoretical interest since it allows one to study the gradual breakdown of the semiclassical diagonal approximation for large trajectory periods. Indeed, we observe significant deviations from the persistent-current quantum results at $k_B T/\Delta \sim 0.25$. In order to study the temperature dependence in more detail, as well as the dependence on the Fermi energy, it is more convenient to consider the averaged magnetic susceptibility $\langle \chi \rangle$ (representing a thermodynamic measure of level curvatures) since the latter is nonzero at zero flux.

4.3.3 Magnetic Susceptibilities

To make contact with the preceding section we first present results for the susceptibility of rings (referring to the system in Fig. 4.4) and then discuss the magnetic response of singly connected chaotic billiards.

Ring Topologies

By taking a second flux derivative one obtains from (4.36) the averaged susceptibility (defined in (4.24b)) for a chaotic cylinder in the semiclassical approximation [70]:

$$\frac{\overline{\chi}}{\chi_L} \simeq \frac{12}{\pi} \frac{a}{w} L_T \, \sigma \sum_{m=-\infty}^{\infty} \int_0^{\infty} d\xi \, R^2(\xi)[1 - 2g_m(\varphi, \xi)] \, \exp\left[-g_m(\varphi, \xi)\right]. \quad (4.40)$$

Here, the g_m are as defined in (4.37) and $\overline{\chi}$ is normalized with respect to the Landau susceptibility $-\chi_L$, (4.7). Figure 4.7 depicts the quantum mechanical (full lines) and semiclassical (dashed lines) results for the flux-dependent average susceptibility of the chaotic cylinder (at different temperature smoothings $k_B T/\Delta = 0.5, 1, 2, 4$). The same ensemble is used as in the previous section. We find reasonable quantitative agreement for the paramagnetic behavior at small fluxes and the crossover to a diamagnetic response, for the different temperatures. The deviations near $\phi = \phi_0/4$ can be traced back to the fact that the semiclassical result at $\phi = \phi_0/4$ stems from the difference of two nearly equal contributions in the integral (4.40), leading to an enhanced relative error.

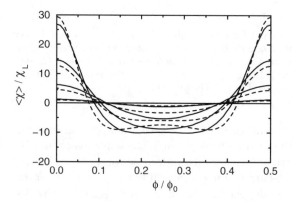

Fig. 4.7. Ensemble-averaged susceptibility $\langle \chi \rangle$ of an ensemble of chaotic ballistic rings (see Fig. 4.4) over one period of the magnetic flux $\varphi = \phi/\phi_0$ for different temperatures ($k_B T/\Delta = 0.5, 1, 2, 4$ from the *top* for the maximum at $\phi = 0$) at $k_F w \simeq 90$. Semiclassical results (*dashed lines*, (4.36)) are compared with exact quantum calculations (*full lines*)

The $\varphi = 1/2$ periodicity of $\langle \chi \rangle$ follows from the "false time reversal symmetry breaking" at half-integer fluxes. The corresponding maximum is semiclassically related to the constructive interference of contributions from periodic orbits and their time reversal counterparts at $\varphi = 1/2$.

The temperature dependence of the line shape can be understood as follows: the smaller the temperature, the larger is L_T and the longer are the typical orbits contributing to $\langle \chi \rangle$. The variance in the distribution of the areas enclosed by these orbits increases, making them more sensitive to the magnetic field and thus yielding a larger susceptibility at zero field and a smaller peak width, since time reversal invariance is more rapidly destroyed.

For $\varphi = 0$ and $L_T \sigma \gg 1/\pi^2$, i.e. in the limit of large k_F, the integral in (4.40) can be solved analytically, giving a linear dependence of $\bar{\chi}$ on L_T and k_F, namely

$$\frac{\overline{\chi(\varphi \equiv 0)}}{\chi_L} \simeq \frac{2}{\pi} \frac{a}{w} \left(\pi^2 L_T \sigma - 2 \right) . \tag{4.41}$$

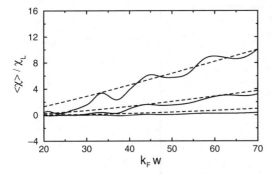

Fig. 4.8. Ensemble-averaged susceptibility $\langle \chi \rangle$ of chaotic ballistic rings as a function of $k_F w$ (k_F is the Fermi momentum, w the width of the structure) at zero flux and normalized to the Landau susceptibility χ_L. The overall linear increase of $\langle \chi \rangle$ obtained from exact quantum mechanical calculations (*full lines*) follows the semiclassical predictions (*dashed lines*, (4.40)) for different temperature smoothings $k_B T/\Delta = 1, 2, 4$ (from the *top*). (From [70], by permission)

Figure 4.8 shows the zero-flux susceptibility as a function of $k_F w$ for different temperatures $k_B T/\Delta = 1, 2$, and 4. The quantum results (full lines) show on the whole a linear increase with $k_F w$, in agreement with the semiclassical prediction (dashed lines) from the equation above. However, even after energy and ensemble averaging, as described above, an oscillatory structure persists in the quantum results. Its origin, which is not yet completely understood, is beyond the present semiclassical treatment. It may be related to the possiblity that the variations in the disk radii in the ensemble used were still too small and off-diagonal products of contributions from different short periodic orbits persist.

As mentioned above, temperature provides a natural parameter for studying smoothing effects on mesoscopic spectral correlations. Figure 4.9 focuses on the temperature dependence: the averaged susceptibility at $\varphi = 0$ and $k_F w \simeq 90$ is shown as a function of the cutoff length $L_T \sim 1/k_B T$ (see (4.39)). The quantum mechanical result (full line) exhibits reasonable agreement with the semiclassical approximation (dashed line) for intermediate cutoff lengths $0.5 \leq L_T/a \leq 4$, which correspond to temperatures of the order $0.7 \leq k_B T/\Delta \leq 5$.

The deviations at *small* L_T indicate the limit of the statistical assumption, the winding-number distribution, used in the semiclassical approach: for $L_T \leq a$ the function $R(\xi)$ in the integral (4.40) exponentially suppresses longer orbits and only a few fundamental periodic orbits will contribute to the magnetic response. Owing to the ring topology the minimum length $L_{\min}$ of flux-enclosing orbits must be larger than a. We find $L_{\min} \simeq 1.5a$ for the present geometry. This allows one to predict the functional dependence of $\langle \chi \rangle$ for high temperatures to be exponential, according to the temperature damping function $R^2(L_{\min}/L_T) \simeq (L_{\min}/L_T)^2 \exp(-2L_{\min}/L_T)$. The inset

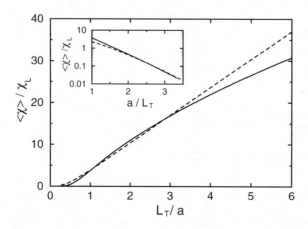

Fig. 4.9. Temperature dependence of the ensemble-averaged susceptibility (at $\varphi = 0$, $k_F w \simeq 90$): $\langle \chi \rangle$ is shown as a function of the thermal cutoff length $L_T \sim \Delta/k_B T$ (see (2.51)) normalized to the circumference a of the cylinder. The semiclassical approximation (*dashed line*) reproduces the quantum mechanical results (*full lines*) reasonably well up to $L_T/a \simeq 4$ or $k_B T/\Delta \simeq 0.7$. At low T signatures from off-diagonal contributions emerge (see text). *Inset*: high-temperature (low-L_T) regime in semilogarithmic representation: the exponent of the quantum results follows precisely a semiclassical estimation (*dashed line*) according to the temperature damping function $R^2(L_{\min}/L_T) \simeq (L_{\min}/L_T)\exp(2L_{\min}/L_T)$. (The semiclassical prefactor (y offset in the log representation) was fitted.) (From [70], by permission)

of Fig. 4.9 depicts semilogarithmically this damping function (dashed line) in comparison with the full quantum results (full line) as a function of a/L_T. The (damping) exponent of the high-temperature susceptibility is well reproduced as $-2L_{\min}/L_T$, confirming the semiclassical picture.

In the case of $L_T/a > 3$ (corresponding to a temperature smoothing $k_B T < \Delta$) the increasing difference between the semiclassical and quantum results in Fig. 4.9 indicates the onset of the breakdown of the diagonal approximation: off-diagonal terms related to pairs of nonidentical periodic orbits, which may exhibit correlations in their actions and stability factors which enter into (4.31), presumably cause the deviation of $\langle \chi \rangle$ from the linear L_T dependence.[4] In this respect, the issue of whether one can find for this system deviations from random-matrix theory at large but finite k_F deserves special interest.

---

[4] Strictly speaking, the relation (4.24b), associating the averaged magnetic response with fluctuations in the particle number, is justified only for $k_B T \geq \Delta$. However, our numerical calculations, comparing the canonical response with that obtained from (4.24b) for a square billiard, and independent work by Mathur et al. [247] show its validity even for temperatures $k_B T \leq \Delta$.

Singly Connected Billiards

For completeness we summarize some results [45] for singly connected chaotic quantum dots in uniform fields: the susceptibility of an individual system is approximated by the field derivatives of $\Delta F^{(1)}$ (see (2.54c)), which again are applied (to leading order in $\hbar$) only to the rapidly varying phases. As a consequence, the second derivative of the contribution of a periodic orbit to $\Delta F^{(1)}$ merely amounts to a multiplication by a factor $(e\mathcal{A}_{\mathrm{po}})^2/(c\hbar)^2$. This gives, for a billiard of area A,

$$\frac{\chi^{(1)}}{\chi_{\mathrm{L}}} = \frac{24\pi m A}{g_{\mathrm{s}}} \sum_{\mathrm{po}} \frac{R(L_{\mathrm{po}}/L_T)}{\tau_{\mathrm{po}}^2} \left(\frac{\mathcal{A}_{\mathrm{po}}}{A}\right)^2 d_{\mathrm{po}}(\mu) , \tag{4.42}$$

where d_{po} is given by (2.27).

The response of an ensemble of structures follows from $\Delta F^{(2)}$ according to (4.23c) and (2.54b). It can be calculated semiclassically as a double sum over all pairs of orbits. The origin of the magnetic *weak-field* response of an ensemble is again a consequence of time reversal symmetry: off-diagonal terms involving an orbit and its time reversal have actions which differ solely by their flux contributions. They survive the averaging process and contribute to $\overline{\chi^{(2)}}$ since they enclose directed areas of opposite sign. One finally obtains, by pairing time-reversal-related terms [45],

$$\frac{\overline{\chi^{(2)}}}{\chi_{\mathrm{L}}} = 24 \sum_{\mathrm{po}} \frac{R^2(L_{\mathrm{po}}/L_T)}{j_{\mathrm{po}}^2 |\det(\boldsymbol{M}_{\mathrm{po}} - \boldsymbol{I})|} \left(\frac{2\mathcal{A}_{\mathrm{po}}}{A}\right)^2 \cos\left(\frac{4\pi\mathcal{A}_{\mathrm{po}}H}{\phi_0}\right) \tag{4.43}$$

with the values of all classical entries taken at $H=0$. At zero field, the cosine in (4.43) for the surviving terms is one and the prefactors are positive. This fact explains the general *paramagnetic* character of the susceptibility of an ensemble. The dephasing of time reversal orbits due to the perturbing magnetic field necessarily induces on average a decrease of the amplitude of the susceptibility. Following equivalent lines to those used for ring geometries, i.e. invoking a Gaussian distribution $P_L(\Theta)$ of enclosed areas, the semiclassical averaged susceptibility of singly connected structures takes the form

$$\frac{\overline{\chi^{(2)}}_{\mathrm{D}}}{\chi_{\mathrm{L}}} = 96\,\Lambda\,\mathsf{F}(\gamma) \tag{4.44}$$

with $\Lambda = \sigma_L L_T / A^2$. The function $\mathsf{F}(\gamma)$ is defined as

$$\mathsf{F}(\gamma) = \int_0^\infty R^2(\xi)\,(1 - 4\gamma^2\xi)\exp(-2\gamma^2\xi)\,\mathrm{d}\xi ; \quad \gamma = \frac{2\pi H}{\phi_0}\sqrt{\sigma_L L_T} . \tag{4.45}$$

Note that this result corresponds to the $m = 0$ component of (4.40). The maximum of $\mathsf{F}(\gamma)$ is $\mathsf{F}(0) = \pi^2/6$ and the half-width $\Delta\gamma \simeq 0.252$. A numerical solution of the quadrature gives a universal line shape curve [45] similar to that depicted in Fig. 4.7 for $\varphi < 1/4$.

4.3.4 Systems with Diffusive Dynamics

For completeness we briefly summarize a semiclassical approach by Argaman, Imry, and Smilansky [76] to persistent currents in *diffusive* mesoscopic
systems[5] and compare it with the above results. These authors express correlation functions for the averaged and typical persistent currents of the type
(4.31) for $T = 0$ in terms of the semiclassical form factor (2.31), the Fourier
transform of the spectral two-point correlator, which they derive to be

$$K(E,t) \simeq \frac{2|t|}{(2\pi\hbar)^2} \frac{d\Omega}{dE} \, P(t) \,. \tag{4.46}$$

$P(t)$ is the classical return probability for periodic motion and $d\Omega/dE$
the volume of the energy shell. For thin rings of perimeter a the dynamics of returning particles obeys a one-dimensional diffusion law $P(na) \sim$
$1/\sqrt{4\pi D|t|} \exp(-n^2 a^2/4D|t|)$, where n denotes the winding number. Further evaluation leads to a final expression for the averaged persistent current
which is equivalent to the expression (4.36) derived for ballistic chaotic systems. This close relation can be traced back to the following fact: although
the return probability for a closed chaotic system is constant ($P(t) = dE/d\Omega$,
contrary to diffusive systems), a one-dimensional diffusion law determines the
accumulation of *areas*, while in the diffusive case it determines the dynamics
itself. Thus, in spite of the different origins of the diffusion processes involved,
the overall results for diffusive one-dimensional rings and chaotic ballistic billiards are of the same form. However, diffusive singly connected systems in
uniform magnetic fields behave differently. There, two-dimensional diffusive
motion determines $P(t)$, *and* the conditional distribution for enclosed areas
is not Gaussian but of the form [76]

$$P_L(\Theta) = \frac{\pi}{2Dt[1 + \cosh(\pi\Theta/Dt)]} \,. \tag{4.47}$$

4.3.5 Relation to Experiments

To our knowledge, two experiments on ballistic mesoscopic structures have
been performed so far: measurements of persistent currents in rings [220] and
of the susceptibility of semiconductor square quantum wells [57]. However,
the classical motion of electrons in both experiments can be considered to be
regular; and these will be addressed in Sects. 4.6.1 and 4.5, where integrable
ring and square billiards are treated.

An experimental observation of the persistent-current or susceptibility
line shapes (see (4.36) and (4.44)) would be desirable as a confirmation of
the applicability of the semiclassical picture developed here (besides the comparison with the quantum results). Although a clear-cut verification of the

---

[5] We note, however that the models of Argaman et al. based on noninteracting
electrons, cannot explain the present experiments on diffusive metal rings (see
Sect. 6.3.2).

precise functional form of the line shape may be too difficult, the characteristic dependence of the magnitude and line shape of the universal orbital
magnetic response functions on temperature should be observable. Since today experiments can be performed at rather low temperatures corresponding
to cutoff lengths of order $L_T/a \simeq 5$ [220], they can access the L_T region
where the statistical assumptions about the behavior of long orbits are fulfilled, leading to universal features in the magnetic response.

However, one should keep in mind that the above quantum and semiclassical analysis holds for noninteracting systems. Interaction contributions to
the magnetic response in chaotic ballistic systems are estimated in Sect. 6.5
and have to be added.

In addition, as will be discussed in detail in Chap. 5, the effect of smooth
disorder has to be considered. This is characteristic of the GaAs/AlGaAs
heterostructures with which this kind of experiment is performed. As a result, an average over weak disorder will actually favor the cancellation of
the nondiagonal terms *without affecting* the diagonal contribution presented
above. The effects of nondiagonal contributions should therefore be less important in actual systems than it might appear in a clean model.[6] However,
for rather clean devices it may even be possible to experimentally approach
the regime of the breakdown of the diagonal approximation where the mesoscopic quantum magnetic response can be expected to be strongly influenced
by correlations between nonidentical classical periodic paths.

Furthermore, new experiments with microwave cavities are in progress
[262] which allow one to study the breaking of time reversal symmetry in "persistent current like" correlation functions of frequency-dependent microwave
absorption spectra. The symmetry breaking is achieved there by using me-
dia which, upon magnetization, induce field-dependent phase shifts in waves
undergoing reflection. These kinds of measurements probe the independent-
particle response and should be sensitive to semiclassical off-diagonal contributions.

4.4 Perturbed Integrable Systems: General Framework

The preceding section was devoted to the magnetic response of chaotic systems, which are assumed to show generic behavior: they exhibit universal
features representing a wide, common class of systems. One example is the
line shape of the weak-field magnetization profile as derived in Sect. 4.3.[7]
In this section we discuss for integrable geometries the generic situation in

---

[6] The variance in the disk radii of the chaotic cylinders in the quantum mechanical ensemble average performed in Sect. 4.3.3 was kept small enough that off-
diagonal terms persisted.

[7] Nevertheless, system-specific features generally coexist in chaotic systems, at
least for properties related to shorter timescales. However, they can be reduced
by an appropriate choice of the model system.

which a perturbation, in our case an external magnetic field, breaks the integrability. We therefore shall refer to the weak-field behavior because only this regime is affected by the integrability of the dynamics at zero field. The symmetry-breaking effect of the magnetic field is therefore twofold: besides the breaking of time reversal symmetry, an additional symmetry of the field-free Hamiltonian (giving rise to its integrable character) is affected.

The aim of the present section is to present the more general implications for the related quantum response on the basis of a *semiclassical perturbation theory*. We shall discuss the consequences for magnetic properties of ballistic quantum dots first on a general level. In the subsequent section we then treat as an application the magnetism of square billiards, which represent experimentally relevant prototypes of generic integrable geometries. The case of systems which remain integrable at arbitrary field strength owing to their special geometry will be discussed in Sect. 4.6.

On the basis of the field derivatives of the thermodynamic functions $\Delta F^{(1)}$ and $\Delta F^{(2)}$, we shall compute in the following the magnetic response of noninteracting, two-dimensional integrable systems perturbed by a weak magnetic field. The equations (2.54b), (2.54c), and (4.23a)–(4.23c), which relate $\Delta F^{(1)}$ and $\Delta F^{(2)}$ to the oscillating part, $d^{\mathrm{osc}}(E)$, of the density of states, are general relations which also apply in particular here. The main difficulty is to obtain semiclassical uniform approximations for $d^{\mathrm{osc}}(E)$ in the perturbative regime: the periodic orbits which play the central role in the semiclassical trace formulas are the most strongly affected by the perturbation. According to the Poincaré–Birkhoff theorem [91] all resonant tori, i.e. all families of periodic orbits, are instantaneously broken as soon as the magnetic field is turned on, leaving only two isolated periodic orbits (one stable and one unstable). It is therefore no longer possible to use the Berry–Tabor trace formula (2.25), suitable for integrable systems, to calculate $d^{\mathrm{osc}}(E, H)$, since it is based on a sum over resonant tori which do not exist any longer. One has therefore to devise a semiclassical technique allowing one to interpolate between the zero-field regime, for which the Berry–Tabor formula applies, and higher fields (still classically perturbative, however), at which the periodic orbits which persist under the perturbation are sufficiently well isolated to use the Gutzwiller trace formula (2.27).

Such an approach is presented in more detail in [45]. It is related to a semiclassical treatment of the density of states in the nearly integrable regime by Ozorio de Almeida [27, 263].[8] Here, we briefly summarize the basic results of this semiclassical treatment in the framework of classical perturbation theory for small fields and then deduce the grand canonical and canonical contributions to the susceptibility.

In the integrable zero-field limit each closed trajectory belongs to a torus I_M and we can replace r in the trace integral (2.18) by angle coordinates

---

[8] A general approach based on that by Ozorio de Almeida was also recently used by Creagh [264].

Θ_1 specifying the trajectory within the one-parameter family and by the position Θ_2 on the trajectory. For small magnetic fields the classical orbits can be treated as being essentially unaffected, while the field acts merely on the phases in the Green function in terms of the magnetic flux through the area $\mathcal{A}_{\mathbf{M}}(\Theta_1)$ enclosed by each orbit of the family $\mathbf{M}$. The evaluation of the trace integral (2.18) along Θ_2 for the semiclassical Green function of an integrable system leads in this approximation to a factorization of the density of states [45]:

$$d^{\mathrm{osc}}(E) = \sum_{\mathbf{M} \neq 0} \mathcal{C}_{\mathbf{M}}(H) d_{\mathbf{M}}^0(E) \ . \tag{4.48}$$

$d_{\mathbf{M}}^0(E)$ stands for the density-of-states contribution (given by the Berry–Tabor formula (2.25)) in the integrable zero-field limit. Furthermore,

$$\mathcal{C}_{\mathbf{M}}(H) = \frac{1}{2\pi} \int_0^{2\pi} d\Theta_1 \, \cos\left[2\pi \frac{H \mathcal{A}_{\mathbf{M}}(\Theta_1)}{\phi_0} \right] \ . \tag{4.49}$$

At zero field we obviously have $\mathcal{C}_{\mathbf{M}}(0) = 1$.

The functions $\mathcal{A}_{\mathbf{M}}(\theta_1)$, and therefore $\mathcal{C}_{\mathbf{M}}(H)$, are system-dependent. Following again Ozorio de Almeida, it is convenient to write $\mathcal{A}_{\mathbf{M}}(\theta_1)$ as a Fourier series

$$\mathcal{A}_{\mathbf{M}} = \sum_{n=0}^{\infty} A_{\mathbf{M}}^{(n)} \sin(n\theta_1 - \gamma^{(n)}) \ . \tag{4.50}$$

If $\mathcal{A}_{\mathbf{M}}$ is a smooth function of θ_1, the coefficients $A_{\mathbf{M}}^{(n)}$ usually rapidly decay with n. For systems where one can neglect all harmonics higher than the first one, the integral (4.49) can be performed. To this end one has to consider two different situations: namely whether the torus is time-reversal-invariant (e.g. a square geometry) or has a partner in phase space which is its counterpart under time reversal transformation (e.g. a circular geometry). Here we note the result for the former case [45]:

$$\mathcal{C}_{\mathbf{M}}(H) \simeq J_0\left(\frac{2\pi H \mathcal{A}_{\mathbf{M}}^{(1)}}{\phi_0} \right) \ . \tag{4.51}$$

As was shown in [45] for the case of the square billiard, representing a generic integrable system, this is an excellent approximation to the exact form (4.64) to be calculated in the next section.

4.4.1 Magnetic Susceptibility

From the expression (4.48) for the oscillating part of the density of states the contributions $\chi^{(1)}$ and $\chi^{(2)}$ to the susceptibility are obtained by the application of (2.54b), (2.54c), and (4.23a)–(4.23c), which express $\Delta F^{(1)}$ and $\Delta F^{(2)}$ in terms of $d^{\mathrm{osc}}(E, H)$. Taking the field derivative twice according to (4.17) and introducing the dimensionless quantities

$$C''_{\mathbf{M}}(H) \equiv \left(\frac{\phi_0}{2\pi A}\right)^2 \frac{\mathrm{d}^2 C_{\mathbf{M}}}{\mathrm{d}H^2} \quad , \quad (C^2)''_{\mathbf{M}}(H) \equiv \left(\frac{\phi_0}{2\pi A}\right)^2 \frac{\mathrm{d}^2 C_{\mathbf{M}}^2}{\mathrm{d}H^2} \qquad (4.52)$$

(A is the total area of the system), one obtains for the grand canonical contribution to the susceptibility

$$\frac{\chi^{(1)}}{\chi_{\mathrm{L}}} = -\frac{24\pi m A}{g_s} \sum_{\mathbf{M}} \frac{R(\tau_{\mathbf{M}}/\tau_T)}{\tau_{\mathbf{M}}^2} d_{\mathbf{M}}^0(\mu) C''_{\mathbf{M}}(H) . \qquad (4.53)$$

$R(x)$ determines the temperature damping on the scale of the time τ_T, (2.50).

Equation (4.53) is the basic equation for the susceptibility of an individual microstructure. As for chaotic systems in the last section, we describe an ensemble of ballistic systems by performing an energy average. Related variations in k_{F} lead, by means of the contributions $k_{\mathrm{F}} L_{\mathbf{M}}$ to the actions $S_{\mathbf{M}}/\hbar$ in a billiard, to large variations ($> 2\pi$) in the phases which enter into (4.53) via the semiclassical expression (2.18) for the density of states. Thus, $\chi^{(1)}$ vanishes upon ensemble averaging. Therefore, the typical susceptibility $\chi^{(t)}$ (see (4.18)) and the averaged susceptibility $\overline{\chi}$ (see (4.24b)) will be considered as relevant measures.

Assuming that there are no degeneracies in the lengths $L_{\mathbf{M}}$ of orbits from different families $\mathbf{M}$, one obtains for $\chi^{(t)}$, using (4.53),

$$\left(\frac{\chi^{(t)}}{\chi_{\mathrm{L}}}\right)^2 = \left(\frac{24\pi}{g_s} m A\right)^2 \sum_{\mathbf{M}} \frac{R^2(\tau_{\mathbf{M}}/\tau_T)}{\tau_{\mathbf{M}}^4} \overline{d_{\mathbf{M}}^0(\mu)^2} \left(\frac{\mathrm{d}^2 C_{\mathbf{M}}}{\mathrm{d}H^2}\right)^2 . \qquad (4.54)$$

In calculating $\overline{\chi}$, the contribution from $\Delta F^{(1)}$ vanishes upon energy averaging. The canonical correction $\Delta F^{(2)}$ in (4.22) gives, in the semiclassical approximation [45],

$$\frac{\overline{\chi^{(2)}}}{\chi_{\mathrm{L}}} = -\frac{24\pi^2 \hbar^2}{g_s^2} \sum_{\mathbf{M}} \frac{R^2(\tau_{\mathbf{M}}/\tau_T)}{\tau_{\mathbf{M}}^2} \overline{[d_{\mathbf{M}}^0(\mu)]^2} (C^2)''_{\mathbf{M}}(H)$$

$$= -\frac{12}{\hbar} \sum_{\mathbf{M}} \frac{R^2(\tau_{\mathbf{M}}/\tau_T)}{M_2^3 |g_\mu''(\boldsymbol{I}_{\mathbf{M}})|} (C^2)''_{\mathbf{M}}(H) . \qquad (4.55)$$

The field-dependent component of $\overline{\chi^{(2)}}$ in the limit $H \to 0$ is given by

$$(C^2)''_{\mathbf{M}}(H=0) = -\frac{1}{2\pi A^2} \int_0^{2\pi} \mathrm{d}\theta_1 \, A_{\mathbf{M}}^2(\theta_1) . \qquad (4.56)$$

It is always negative. Therefore, for an ensemble of integrable structures the magnetic response is always paramagnetic at zero field, in close correspondence to the result (4.43) obtained for an ensemble of chaotic ballistic systems.

4.4.2 Integrable Versus Chaotic Behavior

Let us summarize the similarities and differences in the orbital magnetic response of chaotic and integrable systems with respect to the treatment as well as to the results.

The qualitative behavior of the magnetic response is quite the same for generic chaotic and integrable systems: the susceptibility of a *single* structure can be paramagnetic or diamagnetic and changes sign with a periodicity in $k_F a$ of the order of 2π. The most remarkable similarity is the paramagnetic character of the average susceptibility of an *ensemble* of microstructures. Equations (4.43) and (4.55) show that the susceptibility is positive at zero field, independent of the kind of dynamics considered. Equation (4.23c) states that $\overline{\Delta F^{(2)}}$ is, up to a multiplicative factor, the variance of the temperature-smoothed particle number for a given chemical potential μ. As a basic mechanism for both integrable and chaotic systems, the magnetic field reduces the degree of symmetry of the system, which lowers this variance. Therefore $\overline{\Delta F^{(2)}}$ necessarily decreases when the magnetic field is applied.

For chaotic systems the paramagnetic character of the ensemble susceptibility arises as naturally as the negative sign of the magnetoresistance in phase-coherent microstructures, see Sect. 3.2.3. The reasoning is similar to a random-matrix argument, where the ensembles describing the fluctuations of time-reversal-invariant systems are known to be less rigid: the fluctuations in the number of states in any given interval of energy are larger than in the case where time-reversal invariance is broken. The transition from one symmetry class to the other can be explained on the basis of generalized ensembles, whose validity can be justified semiclassically [84]. We must, however, point out that even for the chaotic case we do not have the standard GOE–GUE transition [23], since (4.23c) involves integration over a large energy interval. For higher temperatures, this characterizes the nonuniversal "saturation" regime where $[N^{\mathrm{osc}}(\mu)]^2$ is given by the shortest periodic orbits.

There are some differences worth considering:

(i)   For chaotic systems the only symmetry existing at zero field is the time reversal invariance, while for integrable systems the loss of time reversal invariance *and* the breaking of invariant tori together reduce the amplitude of $N^{\mathrm{osc}}(\mu)$. This difference arises from the lack of structural stability of integrable systems under a perturbing magnetic field.

(ii) We derived in Sect. 4.3 for chaotic systems at low temperatures, such that a large number of orbits contribute to the susceptibility, universal curves for the field-dependent magnetization similar to the weak-localization effect in electric transport. This is not possible for integrable systems, which do not naturally lend themselves to a statistical treatment.

(iii) The main difference between integrable and chaotic noninteracting systems is related to the *magnitude* of the magnetic response. The contribution of an orbit to the Gutzwiller formula for two-dimensional systems is smaller by $\sqrt{\hbar}$ than a term in the Berry–Tabor formula for the integrable case. More generally, in the case of d degrees of freedom, the $\hbar$ dependence of the Berry–Tabor formula is $\hbar^{-(1+d)/2}$. It is the same as in the semiclassical Green function. The Gutzwiller formula is obtained by performing the trace integral of the Green function by stationary phase in $d-1$ directions, each of which yields a factor $\hbar^{1/2}$. This results in an overall $\hbar^{-1}$ behavior independent of d for a chaotic system. Important consequences therefore arise for the case of two-dimensional billiards of typical size a at temperatures such that only the first few shortest orbits significantly contribute to the free energy. The different $\hbar$ dependence gives rise to different parametric $k_F a$ characteristics of integrable and chaotic systems.

While the magnetic response of chaotic systems results from *isolated* periodic orbits, the existence of *families* of flux-enclosing orbits in quasi-integrable or partly integrable systems is reflected in a parametrically different dependence of their magnetization and susceptibility on $k_F a$ (or $\sqrt{N}$ in terms of the number of electrons). The difference is especially drastic for ensemble averages, where we expect a $k_F a$–independent response $\bar{\chi}$ for a chaotic system, while the averaged susceptibility for integrable systems increases linearly in $k_F a$. This will become more transparent when considering explicitly the example of an ensemble of square potential wells.

A complete comparison between the magnetic responses of integrable and chaotic systems, including the effect of electron–electron interactions, will be given in Chap. 6. There, the different $k_F a$ behaviors of the susceptibility for individual systems as well as ensemble averages is displayed in Table 6.1.

4.5 Perturbed Integrable Systems: Square Quantum Wells

In the preceding section perturbation theory for generic integrable systems was presented and uniform approximate (but general) expressions were derived for the weak-field magnetic response. In this section we treat explicitly the case of a square billiard, which can be considered as the prototype of a generic integrable structure. Because of the simplicity of its geometry, the trace integrals over the Green function can be performed exactly for weak magnetic fields and yield uniform analytical expressions. In order to obtain semiclassical results for the susceptibility of individual and ensembles of squares we proceed as outlined in Sect. 4.2.3: we calculate the density of states and use the decomposition of the susceptibility according to (4.22) into contributions corresponding to the field derivatives of $\Delta F^{(1)}$ and $\Delta F^{(2)}$. The semiclassical computations will be compared with precise numerical quantum calculations.

The square geometry deserves special interest since it was the first microstructure experimentally realized for measuring the magnetic response in the ballistic regime [57]. A semiclassical approach to the magnetism of clean square billiards was first used by Ullmo, Richter, and Jalabert [45, 71] and von Oppen [223]. The treatment in this section assumes clean squares in order to accentuate the physical mechanisms leading to the magnetic response. Effects of residual disorder and electron–electron interactions will be treated in Chaps. 5 and 6. In Sect. 6.6 we are then in a position to compare the results with the experimental findings of [57].

4.5.1 The Weak–Field Density of States

Following [45] closely, we calculate in the first step the oscillating density of states, $d^{\mathrm{osc}}(E)$, from the trace of the semiclassical Green function. To start with, we consider a square billiard of side a in the absence of a field. Each family of periodic orbits can be labeled by the topology $\boldsymbol{M} = (M_x, M_y)$, where M_x and M_y are the numbers of bounces occurring on the bottom and left sides of the billiard (see Fig. 4.10). The length of the periodic orbits for all members of a family is

$$L_{\boldsymbol{M}} = 2a\sqrt{M_x^2 + M_y^2} \ . \tag{4.57}$$

The action along the unperturbed trajectory is $S_{\boldsymbol{M}}^0/\hbar = kL_{\boldsymbol{M}}$ (see (2.9)), where k is the wave number. The Maslov indices are $\eta_{\boldsymbol{M}} = 4(M_x + M_y)$. We shall omit them from now on since they yield only a dephasing of a multiple of 2π.

a)   b)

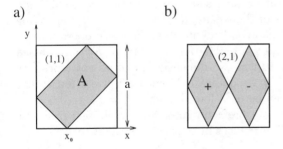

Fig. 4.10. (a) Representative orbits from the family (1,1) of the square billiard. The abscissa x_0 of the intersection of the trajectory with the *lower side* of the square labels the trajectories inside the family. $s \in [0, L_{11}]$ further specifies a particular point of a given trajectory. (b) Orbit of the family (2,1), illustrating the flux cancellation occurring for a whole class of periodic trajectories. (From [45], by permission)

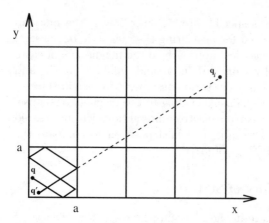

Fig. 4.11. Schematic illustration of the method of images: the Green function $G(\boldsymbol{q}, \boldsymbol{q}')$ is constructed from the free Green function G^0 by placing a source point at each mirror image $\boldsymbol{q}_i$ of the actual source $\boldsymbol{q}$. With each of the resulting mirror images is associated a classical trajectory (*solid line*). This latter is obtained from the straight line joining $\boldsymbol{q}_i$ to $\boldsymbol{q}'$ (*dashed*) by mapping it back onto the original billiard. (From [45], by permission)

It is convenient to use the method of images, as illustrated in Fig. 4.11, to express the exact Green function $G(\boldsymbol{q}, \boldsymbol{q}'; E)$ for a square in terms of the free Green function $G^0(\boldsymbol{q}, \boldsymbol{q}'; E)$ as [79, 90]

$$G(\boldsymbol{q}, \boldsymbol{q}'; E) = G^0(\boldsymbol{q}, \boldsymbol{q}'; E) + \sum_{\boldsymbol{q}_i} \epsilon_i G^0(\boldsymbol{q}_i, \boldsymbol{q}') \ . \tag{4.58}$$

Here the $\boldsymbol{q}_i$ represent all the mirror images of $\boldsymbol{q}$ obtained by any combination of symmetries across the sides of the square, and $\epsilon_i = +1$ or -1, depending on whether one needs an even or odd number of symmetry operations to map $\boldsymbol{q}$ to $\boldsymbol{q}_i$. The semiclassical approximation to $G^0(\boldsymbol{q}, \boldsymbol{q}'; E)$ is discussed in Sect. 2.1.1 and given in (2.10); in two dimensions, it does not reproduce the logarithmic behavior, for $\boldsymbol{q} \to \boldsymbol{q}'$ but exhibits the correct long-range asymptotic behavior, which can be used for all images $\boldsymbol{q}_i \neq \boldsymbol{q}'$.

The determinant entering as a prefactor into the contribution from one trajectory to the Green function reduces in the zero-field limit to

$$D_{\boldsymbol{M}} = \frac{m}{\sqrt{\hbar k L_{\boldsymbol{M}}}} \ . \tag{4.59}$$

In the limit of weak magnetic fields, we make use of the same approach as in the previous sections, keeping in (2.3) the zeroth-order approximation of the prefactor $D_{\boldsymbol{M}}$ and using the first-order correction δS to the action, which is proportional to the area enclosed by the unperturbed trajectory. As in the generic case the enclosed area varies within a family, contrary to the circular or annular geometries to be discussed in Sect. 4.6.

The contribution to the density of states of the family of recurrent trajectories, which for $H \to 0$ tends to the family of the shortest periodic orbits with nonzero enclosed area, plays a dominant role in determining the magnetic response, as already recognized in [57]. For $H = 0$, this family consists in the set of orbits shown in Fig. 4.10a, which hit the bottom of the square at $0 \le x_0 \le a$. As configuration space coordinates we use the distance x_0 which labels the trajectory, the distance s along the trajectory, and the index $\epsilon = \pm 1$ which specifies the direction in which the trajectory is traversed. In this way, each point $\boldsymbol{q}$ is counted four times, corresponding to the four sheets of the invariant torus. The enclosed area $\mathcal{A}_\epsilon(x_0, s)$ is independent of s. It reads

$$\mathcal{A}_\epsilon(x_0) = \epsilon\, 2\, x_0\, (a - x_0)\,. \tag{4.60}$$

The contribution of the family $(1,1)$ to $d^{\mathrm{osc}}(E)$ is $d_{11}(E) = -(g_{\mathrm{s}}/\pi)\cdot$ Im $\mathcal{G}_{11}(E)$. Inserting (4.60) and (2.15) into the integral of (2.18) one finds [71]

$$\mathcal{G}_{11}(H) = \frac{1}{\mathrm{i}\hbar}\frac{1}{\sqrt{2\mathrm{i}\pi\hbar}} \tag{4.61}$$

$$\times \int_0^{L_{11}} \mathrm{d}s \left(\frac{\mathrm{d}y}{\mathrm{d}s}\right)\int_0^a \mathrm{d}x_0 \sum_{\epsilon = \pm 1} D_{11} \exp\left[\mathrm{i}kL_{11} + \mathrm{i}\frac{2e\epsilon}{\hbar c}Hx_0(a - x_0)\right].$$

As shown in the previous section for the general case, the contribution to the density of states of the family $(1,1)$ factorizes into an unperturbed (Berry–Tabor-like) term and a field-dependent factor:

$$d_{11}(E, H) = d_{11}^0(E)\, \mathcal{C}(H)\,. \tag{4.62}$$

For the square these factors read

$$d_{11}^0(E) \equiv d_{11}(E, H = 0) = \frac{4g_{\mathrm{s}}}{\pi}\frac{ma^2}{\hbar^2(2\pi kL_{11})^{1/2}}\, \sin\left(kL_{11} + \frac{\pi}{4}\right) \tag{4.63}$$

and

$$\mathcal{C}(H) = \frac{1}{a}\int_0^a \mathrm{d}x_0 \cos\left(\frac{2e}{\hbar c}Hx_0(a - x_0)\right)$$

$$= \frac{1}{\sqrt{2\varphi}}\left[\cos(\pi\varphi)\mathrm{C}(\sqrt{\pi\varphi}) + \sin(\pi\varphi)\mathrm{S}(\sqrt{\pi\varphi})\right]\,. \tag{4.64}$$

Here

$$\varphi = \frac{Ha^2}{\phi_0} \tag{4.65}$$

is the total flux through the square measured in units of the flux quantum. In (4.64) C and S stand for the cosine and sine Fresnel integrals [265].

The field dependence of d^{osc} in (4.64) reflects the detuning of phases between time-reversal families of orbits and the field-induced decoherence of different orbits *within* a given family. When φ is of the order of one, the

Fresnel integrals in (4.64) can be replaced by their asymptotic value $1/2$. This amounts to evaluating the integral $\mathcal{C}(\varphi)$ using stationary phase and yields [71]

$$\mathcal{C}^{\mathrm{S}}(\varphi) = \frac{\cos(\pi\varphi - \pi/4)}{\sqrt{4\varphi}} . \tag{4.66}$$

For any nonzero field only the two trajectories corresponding to $x_0 = a/2$ remain periodic (one stable, one unstable, according to the two possible directions of traversal). For $\varphi > 1$ the dominant contribution to $\mathcal{C}(\varphi)$ comes from the neighborhood of these two surviving periodic orbits, and the oscillations of $\mathcal{C}(\varphi)$ are related to the successive dephasing and rephasing of these orbits. Indeed, using the Gutzwiller trace formula with a first-order classical perturbative evaluation of the actions and stability matrices just gives $\mathcal{C}^{\mathrm{S}}(\varphi)$. It, however, diverges when $H \to 0$, while the full expression (4.64) reduces to $\mathcal{C}(0) = 1$.

In order to take into account the contribution d_M of longer trajectories, we write (M_x, M_y) as (ju_x, ju_y), where u_x and u_y are coprime integers labeling the primitive orbits and j is, as usual, the number of repetitions. The effective algebraic area enclosed by a long periodic trajectory in the square is small owing to cancellations. Simple geometry shows that, keeping x_0 as a label of the orbit (with $x_0 \in [0, a/u_x]$ to avoid double counting), the total area enclosed by the trajectory (ju_x, ju_y) is

$$\mathcal{A}_M = \begin{cases} 0 & u_x \text{ or } u_y \text{ even} \\ j\dfrac{\mathcal{A}_\epsilon(u_x x_0)}{u_x u_y} & u_x \text{ and } u_y \text{ odd} \end{cases} . \tag{4.67}$$

Here, $\mathcal{A}_\epsilon(x_0)$ is given by (4.60). Using the above equation, the density-of-states contribution (4.62) for the orbit $(1,1)$ can be generalized to orbits of topology M and reads

$$d_M(E, H) = d_M^0(E)\, \mathcal{C}_M(\varphi) , \tag{4.68}$$

where

$$\mathcal{C}_M(\varphi) = \begin{cases} 1 & u_x \text{ or } u_y \text{ even} \\ \mathcal{C}\left(\dfrac{r\varphi}{u_x u_y}\right) & u_x \text{ and } u_y \text{ odd} \end{cases} . \tag{4.69}$$

Here, $\mathcal{C}(\varphi)$ is as defined in (4.64) and $d_M^0 \equiv d_M(H = 0)$ is given by (4.63), but with L_{11} replaced by L_M.

4.5.2 Susceptibility of Individual Samples and Ensemble Averages

Semiclassical Results

On the basis of the semiclassical expressions (4.62)–(4.64) for the density of states, we shall compute the magnetic response of square quantum dots and compare it in the next subsection with corresponding results from numerically exact quantum calculations.

We begin with the calculation of the semiclassical susceptibility contribution of the family $(1,1)$ of the shortest flux-enclosing orbits only. This corresponds to the temperature regime of the experiment [57], where the characteristic length L_T (2.49) is of the order of L_{11}. In Sect. 4.5.2 we shall generalize this to arbitrary temperature by taking into account the contribution of longer orbits.

Using the expressions (4.62) and (4.63) for the contributions of the family $(1,1)$ to $d^{\mathrm{osc}}(E, H)$, one can compute the corresponding contribution to $\Delta F^{(1)}$ (see (2.54c) and (4.23b)). Taking the derivatives with respect to the magnetic field, we obtain from $\Delta F^{(1)}$ for $L_T \simeq L_{11}$ [45]

$$\frac{\chi^{(1)}}{\chi^0} = \sin\left(k_{\mathrm{F}}L_{11} + \frac{\pi}{4}\right)\int_0^a \frac{dx_0}{a}\mathcal{A}^2(x_0)\cos[\varphi\mathcal{A}(x_0)] \qquad (4.70)$$

with $\mathcal{A}(x_0) = 4\pi x_0(a - x_0)/a^2$ and

$$\chi^0 = \chi_{\mathrm{L}}\frac{3}{(\sqrt{2}\pi)^{5/2}}(k_{\mathrm{F}}a)^{3/2}R(L_{11}/L_T). \qquad (4.71)$$

$R(L_{11}/L_T)$ is the temperature-dependent reduction factor of (2.49).

The susceptibility of an individual square displays pronounced oscillations as a function of the Fermi energy and can be paramagnetic or diamagnetic, as shown Fig. 4.12a. Since we use only one kind of trajectory the typical susceptibility $\chi^{(t)}$ (4.18) is proportional to the prefactor of $\chi^{(1)}$ and hence of the order of $(k_{\mathrm{F}}a)^{3/2}\chi_{\mathrm{L}}$. This is much larger than the Landau susceptibility χ_{L}.

The susceptibility oscillations have a similar origin to the shell effects [30] in metal clusters. Indeed, related oscillations in the magnetic moment of individual transition-metal dots have been experimentally observed [224] and theoretically explained within a semiclassical shell model [266].

Figure 4.12b shows (solid line) that $\chi^{(1)}$ also exhibits oscillations as a function of the flux at a given number of electrons in the square. The susceptibility obtained from $\mathcal{C}^{\mathrm{S}}$ (dashed line) diverges for small flux but proves to be a good description of $\chi^{(1)}$ for $\varphi \overset{>}{\sim} 1$.

For an ensemble of squares of different sizes a, $\chi^{(1)}$ vanishes under averaging if the dispersion of $k_{\mathrm{F}}L_{11}$ across the ensemble is larger than 2π. Then the average susceptibility is given by the contribution to $\Delta F^{(2)}$ from the $(1, 1)$ family, (4.23c). Proceeding in a similar way to that for the first-order term, the contribution of the family $(1, 1)$ to the integrated density N^{osc} is given by (2.54b) as

$$N_{11}(\bar{\mu}, \varphi) = -g_{\mathrm{s}}\left(\frac{2^3a^3}{\pi^3 L_{11}^3}\right)^{1/2}(k_{\mathrm{F}}a)^{1/2} \qquad (4.72)$$

$$\times \cos\left(k_{\mathrm{F}}L_{11} + \frac{\pi}{4}\right)\mathcal{C}(\varphi)\,R\left(\frac{L_{11}}{L_T}\right).$$

(a)

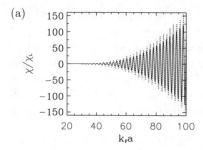

(b)

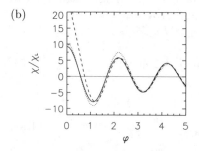

Fig. 4.12. Orbital magnetic response of a single square billiard. (a) χ as a function of $k_F a$ from numerical calculations (*dotted line*) at zero field and at a temperature equal to 10 level spacings. The number of electrons is $N = g_s(k_F a)^2/(4\pi)$. The *full line* shows the semiclassical approximation (4.70) taking into account only the family (1,1) of the shortest orbits. The period $\pi/\sqrt{2}$ of the quantum result indicates the dominance of the shortest periodic orbits enclosing nonzero area with length $L_{11} = 2\sqrt{2}a$. (b) χ as a function of the normalized flux through the sample (at a Fermi energy corresponding to $N \sim 400$) from (4.70) (*solid line*) and from numerical calculation (*dotted line*). The susceptibility arising from the stationary-phase integration $\mathcal{C}^S$ (4.66), shown as the *dashed line* diverges as $\varphi \to 0$. (From [45], by permission)

In order to compute $\chi^{(2)}$ we have to evaluate $\Delta F^{(2)} = (N^{osc})^2/2\bar{D}$, with $\bar{D} = (g_s m a^2)/(2\pi\hbar^2)$. In particular, the term from the family $(1,1)$ reads

$$\frac{[N_{11}(\bar{\mu}, \varphi)]^2}{2\bar{D}} = \frac{g_s \hbar^2 (k_F a)}{(\sqrt{2})^3 \pi^2 m a^2} \cos^2\left(k_F L_{11} + \frac{\pi}{4}\right) \mathcal{C}^2(\varphi)\, R^2\left(\frac{L_{11}}{L_T}\right). \quad (4.73)$$

The above term is of lower order in $k_F a$ than $\Delta F^{(1)}$. However, its sign does not change as a function of the phase $k_F L_{11}$, and hence the square of the cosine survives the ensemble averaging.[9] After taking the derivatives with respect to φ one finds (in the regime $L_T \simeq L_{11}$) [45]

$$\frac{\overline{\chi^{(2)}}}{\chi_L} = -\frac{3}{(\sqrt{2}\pi)^3} k_F a \frac{d^2 \mathcal{C}^2}{d\varphi^2} R^2(L_{11}/L_T). \quad (4.74)$$

---

[9] Possible contributions from the orbits $(1,0)$ and $(0,1)$, which are even shorter than $(1,1)$, are discussed in Sect. 4.5.3.

The overall averaged susceptibility of an ensemble of noninteracting squares is therefore

$$\overline{\chi} = -\chi_{\mathrm{L}} + \overline{\chi^{(2)}} \ .$$

Here, the diamagnetic (bulk) Landau contribution, $-\chi_{\mathrm{L}}$, arising from $\hbar$ corrections to F^0, has been added. In the regime $L \simeq L_T$ that we are considering here, χ_{L} is negligible with respect to $\overline{\chi^{(2)}}$ as $\hbar \to 0$. Hence one can approximate $\overline{\chi} \simeq \overline{\chi^{(2)}}$. In the limit $L_T \ll L$, (4.70) and (4.74) remain valid, but $\chi^{(1)}$ and $\chi^{(2)}$ are exponentially suppressed. In this case χ and $\overline{\chi}$ reduce to the Landau susceptibility, which is independent of the underlying classical dynamics.

The field-dependent function $\mathcal{C}$ has its absolute maximum at $\varphi=0$. Hence the average zero-field susceptibility is *paramagnetic* with a maximum value of [71, 223]

$$\overline{\chi^{(2)}}(\varphi=0) = \frac{4\sqrt{2}}{5\pi} \ k_{\mathrm{F}}a \ \chi_{\mathrm{L}} \ R^2(L_{11}/L_T) \ . \tag{4.75}$$

The linear dependence of the average susceptibility on k_{F} is shown in Fig. 4.13a. For small fields the average susceptibility (thin solid line, Fig. 4.13b) decays on the whole as $1/\varphi$ and additionally oscillates on the scale of one flux quantum through the sample. The period of the field oscillations of the ensemble average is half of that of the corresponding individual systems (see Fig. 4.12b), similarly to the disordered case [214]. This behavior can be traced to the $\mathcal{C}^2$ dependence that appears in (4.74), in contrast to the simple $\mathcal{C}$ dependence of (4.70).

With regard to the experiment of [57] representing an ensemble with a wide distribution of lengths, an average $\langle \cdots \rangle_a$ on a classical scale (i.e. $\Delta a/a \not\ll 1$) rather than on a quantum scale ($\Delta(k_{\mathrm{F}}a) \simeq 2\pi$) is required. Hence the dependence of $\mathcal{C}$ on a through φ has additionally to be considered. Since the scale of variation of $\mathcal{C}$ with a is much slower than that of $\sin^2(k_{\mathrm{F}}L_{11})$ we can effectively separate the two averages. The total mean is introduced by averaging the local mean as

$$\langle \chi \rangle_a = \int \mathrm{d}a \ \overline{\chi} \ P(a) \ . \tag{4.76}$$

Here, the quantum average $\overline{\chi}$ is given by (4.74) and $P(a)$ is the probability distribution of sizes a. For a Gaussian distribution $P(a)$ with a 30% dispersion one obtains the thick solid line of Fig. 4.13b. While the zero-field behavior remains unchanged, the low-field oscillations with respect to φ are suppressed under the second average.

Comparison with Quantum Mechanical Results

In order to check the validity of the semiclassical predictions for the susceptibility, precise quantum calculations have been performed [71] based on a

(a)

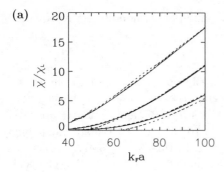

(b)

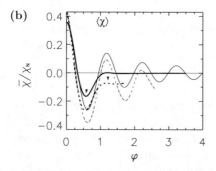

Fig. 4.13. Magnetic susceptibility of an ensemble of squares of different sizes. **(a)** Averaged susceptibility as a function of $k_F a$ for three temperatures ($k_B T/\Delta = 8, 6,$ and 4 for the three triplets of curves, from *bottom* to *top*) and a flux $\varphi = 0.15$. *Dashed curves*: average of the numerically calculated full canonical susceptibility. *Solid* and *dotted curves*: average of $\chi^{(2)}$ calculated semiclassically according to (4.74) and numerically, respectively. **(b)** Flux dependence of the susceptibility (normalized to $\chi_N = \chi_L k_F a R_T^2(L_{11})$) at $k_F a \simeq 70$ from the semiclassical expression (4.74) (*solid*) and numerical calculations (*dashed*). The *thick solid* and *dashed curves* denote averages of the semiclassical and numerical results, respectively, over an ensemble with a large dispersion of sizes, which is denoted by $\langle \chi \rangle$ (see text). The shift of the numerical results with respect to the semiclassical results reflects the Landau susceptibility and effects from bouncing-ball orbits (see Sect. 4.5.3) not included in the semiclassical trace. (From [45], by permission)

diagonalization of the Hamiltonian of noninteracting electrons in a square quantum well subject to a uniform magnetic field (see Appendix A.2).

A typical energy-level diagram of one symmetry class ($(P_\pi, P_{\pi/2}) = (1,1)$) as a function of the magnetic field is shown in Fig. 4.1 (Sect. 4.1). In between the two separable limiting cases $\varphi = 0$ and $\varphi \longrightarrow \infty$, where the levels merge into Landau states, the spectrum exhibits a complex structure typical of a nonintegrable system.

From the single-particle energies obtained we calculated both the grand canonical and the canonical orbital magnetic response without relying on the free-energy expansion (4.22). As an example the numerically calculated susceptibility of a single square well for a given Fermi energy is depicted (as the dotted line) in Fig. 4.2. The pronounced oscillations at intermediate and strong flux will be analysed in Sect. 4.5.3. Here we focus on the magnetic response at weak fields.

The numerical results for the susceptibility of individual and ensembles of squares are displayed as the dashed lines in Figs. 4.12 and 4.13. These results are in excellent agreement with the semiclassical predictions of Sect. 4.5.2. Figure 4.12a shows the numerical quantum result for the canonical susceptibility and the semiclassical leading-order contribution, $\chi_{11}^{(1)}$, at zero field as a function of $k_F a$ (or $\sqrt{4\pi N / g_s}$ in terms of the number of electrons). As expected semiclassically (see (4.70)), the quantum result oscillates with a period $\pi/\sqrt{2}$. This behavior indicates the pronounced effect of the fundamental orbits of length $L_{11} = 2\sqrt{2}a$. The semiclassical amplitudes (solid line) are slightly smaller than the numerical results because only the shortest orbits are considered.

The flux dependence of χ for a fixed number of electrons, $N \approx 1100g_s$, is displayed in Fig. 4.12b. The semiclassical prediction ((4.70), solid curve) is again in considerable agreement with the quantum result, while the analytical result ((4.66), dashed line) from stationary-phase integration yields an (unphysical) divergence for $\varphi \to 0$ as discussed in Sect. 4.5.1. For the numerical calculations the size averages on the quantum scale (thin dashed line, Fig. 4.13b) and the classical scale (thick dashed line) were obtained by taking a Gaussian distribution of sizes with a small or large $\Delta a/a$ dispersion, respectively. Figure 4.13a depicts the $k_F a$ dependence of $\overline{\chi}$ assuming a Gaussian distribution of lengths a with a standard deviation $\Delta a/a \approx 0.1$ for each of the three temperatures $k_B T/\Delta = 2, 3, 5$. The dashed curves are the ensemble averages of the quantum mechanically calculated *entire* canonical susceptibility $\overline{\chi}$. The dotted lines are the *exact* (numerical) results for the averaged term $\overline{\chi_{qm}^{(2)}} = \overline{(N^{osc}{}_{qm})^2}/2\Delta$. They are nearly indistinguishable (on the scale of the figure) from the *semiclassical* approximation of (4.74) (solid line). The precision of the semiclassical approximation based on the fundamental orbits (1,1) is striking. The difference between the results for $\overline{\chi}$ and $\overline{\chi^{(2)}}$ allows one to estimate the precision of the thermodynamic expansion (4.22) used. The semiclassical result for the average on the classical scale has been shifted additionally by $-\chi_L$ to account for the Landau diamagnetism. It is again in close agreement with the numerical results for the averaged susceptibility $\overline{\chi}$.

Lévy et al. [57] measured the orbital magnetic response of an array of 10^5 microscopic quantum billiards of square geometry on a high-mobility GaAs heterojunction. They observed a huge paramagnetic peak of the magnetic susceptibility at zero field, decreasing on a scale of approximately one flux quantum through each square (see Chap. 1, Fig. 1.5). The peak maximum was of the order of $\approx 100\chi_L$. Although our theoretical results for clean square wells with noninteracting particles also give a paramagnetic response with a peak height in reasonable agreement with experiment at low temperatures, we shall give a more detailed comparison with experiment after having considered disorder and interaction effects in Chaps. 5 and 6.

Contribution of Longer Orbits

Although the above comparison with quantum calculations have already shown that at finite temperatures orbits of the family $(1,1)$ yield the dominant contribution to the susceptibility, we discuss here the effect of longer recurrent orbits. In the low-temperature limit or, more generally, if one is interested in results valid at any temperature, it is a priori necessary to take them also into account. Following exactly the same lines as for the contribution of the family $(1,1)$, one obtains from (4.68) the contribution of the family $\boldsymbol{M} = (M_x, M_y) = (ju_x, ju_y)$ (where u_x and u_y are coprime) to $\Delta F^{(1)}$:

$$
\Delta F_{\boldsymbol{M}}^{(1)}(H) = \frac{g_s \hbar^2}{m} \left(\frac{2^3 a}{\pi^3 L_M^5} \right)^{1/2} (k_F a)^{3/2}
$$

$$
\times \sin\left(k_F L_M + \frac{\pi}{4} \right) \mathcal{C}_{\boldsymbol{M}}(\varphi) R\left(\frac{L_M}{L_T} \right) ,
\qquad (4.77)
$$

where $\mathcal{C}_{\boldsymbol{M}}(\varphi)$ is as defined in (4.69). L_M and the function $\mathcal{C}(\varphi)$ are given by (4.57) and (4.64), respectively. In order to find $\chi^{(1)}$ we have to take the second derivative of $\mathcal{C}_{\boldsymbol{M}}$ with respect to the magnetic field. This gives zero if either u_x or u_y is even and a factor $j^2/(u_x u_y)^2$ if both are odd. The result then reads [45]

$$
\frac{\chi^{(1)}}{\chi_L} = -\frac{3}{\pi^{5/2}} (k_F a)^{3/2}
\qquad (4.78)
$$

$$
\times \sum_j \sum_{u_x, u_y} \frac{\sin(k_F L_M + \pi/4)}{j^{1/2} (u_x^2 + u_y^2)^{5/4} (u_x u_y)^2} \, \mathcal{C}''\left(\frac{j\varphi}{u_x u_y} \right) R\left(\frac{L_M}{L_T} \right) ,
$$

$(u_x, u_y$ odd), which is valid at arbitrary temperature. The corresponding low-temperature result for $\chi^{(2)}$ follows essentially along the same lines. As shown in [45], one obtains for the canonical correction to the susceptibility

$$
\frac{\overline{\chi^{(2)}}}{\chi_L} = -\frac{3}{\pi^3} k_F a
$$

$$
\times \sum_j \sum_{u_x, u_y} \frac{R^2(L_M/L_T)}{j \left[(u_x/e)^2 + (u_y e)^2 \right]^{3/2} (u_x u_y)^2} (\mathcal{C}^2)''\left(\frac{j\varphi}{u_x u_y} \right)
\qquad (4.79)
$$

with u_x, u_y odd. The average is taken over rectangles of horizontal and vertical lengths ae and a/e to exclude the possibility of degeneracies in the lengths $L_M = 2a\sqrt{(M_x/e)^2 + (M_y e)^2}$ of different orbits, which may appear in the special case of a square.

Inspection of (4.78) and (4.79) shows that even at zero temperature the strong flux cancellation, which is typical for rectangular geometries, results in a tiny prefactor. For instance, for the second shortest contributing prime orbit, $\boldsymbol{M} = (1, 3)$, one finds for $\chi^{(1)}$ a damping of $1/(9 \times 10^{5/4}) \simeq 0.0062$. For $\overline{\chi^{(2)}}$ the multiplicative factor is even smaller. In practice only the repetitions (jj) of the family (11) will contribute significantly to the susceptibility, and one can use (4.78) and (4.79) keeping only the term $u_x = u_y = 1$ of the second summation. As a result, any possible complications due to the degeneracies in the orbits' lengths for the square are of no practical relavance, and (4.79) restricted to $u_x = u_y = 1$ can be used for the square with $e = 1$. As depicted in Fig. 4.14 for $\overline{\chi^{(2)}}$, the repetitions of the orbit (11) lead to a diverging susceptibility at zero field when the temperature goes to zero. However, they barely affect the result for finite H, even as $T \to 0$, since then the contributions of higher repetitions no longer add coherently.

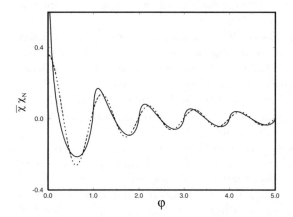

Fig. 4.14. Low-temperature limit of the average susceptibility $\overline{\chi^{(2)}}$ of an ensemble of squares, shown as the *solid line* and given by (4.79). *Dashed curve*: contribution of the family (1,1) to this result. (From [45], by permission)

4.5.3 Bouncing-Ball Magnetism

So far we have discussed effects due to a weak magnetic field such that the classical cyclotron radius r_{cyc} is large compared to the typical size a of the system: $r_{\mathrm{cyc}}/a = c\hbar k/(e\,H\,a) \gg 1$. In that regime, the bending of the electron trajectories due to the magnetic field can be neglected and the main effect of the magnetic field is a change in the semiclassical actions according to the enclosed flux. This semiclassical perturbative approach is valid up to much higher magnetic fields than a corresponding quantum mechanical perturbation theory, as visible in Fig. 4.2 in Sect. 4.2.3. First-order quantum perturbation theory is typically valid up to magnetic fluxes where the first avoided level crossings arise. The figure shows that low-field oscillations of

χ for a single square are well described by semiclassical perturbation theory using unperturbed orbits of the family (11) (left inset in Fig. 4.2) up to field strengths $\varphi \approx 10$. This is orders of magnitude larger than the flux scale relevant to the breakdown of the quantum perturbative approaches. Since the "small" classical parameter is $a/r_{\text{cyc}} \sim H/k_{\text{F}}$, the semiclassical regime of "weak" fields grows with increasing Fermi energy.

On the basis of recent work [45] we present in the following an approach which goes beyond the classically perturbative regime and includes the bending of the classical orbits at larger fields. In that regime, orbital magnetism reflects quantum mechanically the interplay between the scale of the confining energy and the scale of the magnetic-field energy $\hbar\omega_{\text{cyc}}$.

Figure 4.2 depicts a whole scan of the magnetic susceptibility for the square geometry up to a flux $\varphi = 55$, corresponding to $3r_{\text{cyc}} \approx a$. Three different field regimes can be distinguished: weak ($a \ll r_{\text{cyc}}$), intermediate ($a \simeq r_{\text{cyc}}$), and high ($a \geq 2r_{\text{cyc}}$) fields.

The *high-field* regime is classically characterized dominantly by orbits which perform cyclotron motion as long as they do not hit the boundary. One finds the well known de Haas–van Alphen oscillations as in the bulk susceptibility. However, as we shall show, the destruction of part of the cyclotron orbits due to reflections at the boundaries can be semiclassically taken into account. By that means one can describe the crossover regime where $2r_{\text{cyc}}/a$ is smaller than one but not yet negligible.

While the classical dynamics at high field is in general quasi-integrable, the classical phase space in the *intermediate-field* regime is always mixed: Both chaotic and regular motion coexist. Only particular systems with rotational symmetry, which remain integrable independent of the magnetic field, (Sect. 4.6) are an exception. Contrary to the weak-field regime, the intermediate-field regime is characterized by the complete loss of time reversal symmetry. As demonstrated in Fig. 4.2 there appear – besides the weak-field oscillations due to orbits (1,1) – pronounced susceptibility oscillations in the intermediate field regime ($2r_{\text{cyc}} > a$). These reflect quantized *bouncing-ball periodic orbits* (second inset), periodic electron motion due to reflection between opposite boundaries. These oscillations are specific to structures for which the boundary contains at least pieces of parallel straight lines.

We shall review results for this field regime quantitatively again for the case of square microstructures, although the results to be reported are of quite general nature. We refer to individual squares and hence work within the grand canonical formalism.

The quantum mechanically calculated susceptibility of a single square billiard with $\sim$2100 enclosed electrons is shown as the full line in Fig. 4.15a for small and intermediate fluxes at a temperature such that $k_{\text{B}}T/\Delta = 8$. The corresponding semiclassical result $\chi_{(11)}^{(1)}$ from the family (1,1) (see (4.70)) is depicted (with negative offset) as the dashed–dotted line in Fig. 4.15a. Deviations from the quantum result with respect to phase and amplitude begin

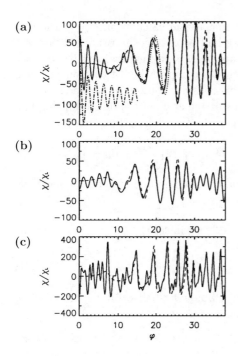

Fig. 4.15. Susceptibility of a single square potential well as a function of magnetic flux. The *full lines* always denote the quantum mechanical results. (a) χ/χ_L calculated at a Fermi energy of 2140 enclosed electrons at a temperature $kT/\Delta = 8$. *Dashed and dotted* lines: semiclassical result due to bouncing-ball orbits from (4.86) with action S_{10} according to (4.81) and (4.87). *Dashed–dotted line*: susceptibility contribution from family (1,1) from (4.70) with offset of -80 for reasons of representation. (b) As (a) but for 1440 electrons and $kT/\Delta = 7$. The semiclassical result (4.88) from bouncing-ball orbits is shown as the *dashed line*. (c) As (b) but for temperature $kT/\Delta = 2$. (From [45], by permission)

to appear at $\varphi \approx 8$ ($r_{cyc} \approx 2a$). They indicate the breakdown of the straight-line approximation for the orbits included. With further increase of the flux a regime is reached where the nonintegrability of the quantum system manifests itself in a complex-structured energy-level diagram (see Fig. 4.1). On the classical level this corresponds to a mixed classical phase space [274] characterized by a variety of coexisting isolated stable and unstable periodic orbits. In particular, there remains a family of orbits with specular reflections on opposite sides of the square only. They are known as "bouncing-ball" orbits in the theory of billiards without a magnetic field. According to the coding introduced in Sect. 4.5.1 they are suitably labeled by the numbers of bounces $(M_x, 0)$ and $(0, M_y)$ at the bottom and left sides of the square, respectively. Since these orbits form families one expects that their corresponding susceptibility contribution should exhibit a parametric dependence on $k_F a$, strongly dominating the contributions of coexisting isolated periodic orbits.

To begin with, we summarize a semiclassical calculation [45] of the susceptibility contribution from bouncing-ball orbits for the primitive periodic orbits $(M_x, 0) = (1, 0)$. The results are then easily generalized to the case of arbitrary repetitions. We start with the computation of the semiclassical Green functions, where we now have to consider the full field dependence of the classical motion. A recurrent path starting at a point q on a bouncing-ball orbit contributes to the diagonal part of the Green function through

$$\mathcal{G}_{10}(q, q' = q; E, H) = \frac{1}{i\hbar\sqrt{2\pi i\hbar}} D_{10} \exp\left[i\left(\frac{S_{10}}{\hbar} - \eta_{10}\frac{\pi}{2}\right)\right] . \quad (4.80)$$

The classical entries such as length, enclosed area, and action follow from simple geometry:

$$L_{10}(H) = \frac{2a\zeta}{\sin\zeta} \quad ; \quad A_{10}(H) = -(2\zeta - \sin 2\zeta)\, r_{\text{cyc}}^2 ;$$

$$\frac{S_{10}}{\hbar} = k\left(L_{10} + \frac{A_{10}(H)}{r_{\text{cyc}}(H)}\right) . \quad (4.81)$$

The angle ζ between the tangent to a bouncing-ball orbit at the point of reflection and the normal to the side is given by

$$\sin\zeta = \frac{a}{2r_{\text{cyc}}} . \quad (4.82)$$

The Maslov index η_{10} is 4 and will therefore be omitted. The prefactor D_{10} was derived in [45]:

$$D_{10}(q, q' = q) = \frac{1}{v}\sqrt{\frac{\hbar k \cos\zeta}{2a}} \quad (4.83)$$

with $v = \hbar k / m$. For the contribution of the whole family $(1, 0)$ we must perform the trace integral (2.18), and obtain for the bouncing-ball contribution to the density of states $d_{10} = -(g_s/\pi)\,\text{Im}\,\mathcal{G}_{10}$ [45]

$$d_{10}(E, H) = -\frac{2g_s}{(2\pi\hbar)^{3/2}} l(H) L_{10} D_{10} \sin\left(\frac{S_{10}}{\hbar} + \frac{\pi}{4}\right) , \quad (4.84)$$

where the length factor

$$l(H) = a\left(1 - \tan\frac{\zeta}{2}\right) \quad (4.85)$$

denotes the field-dependent effective range for the lower reflection points of bouncing-ball trajectories $(1, 0)$. $l(H)$ vanishes for magnetic fields corresponding to $2r_{\text{cyc}} = a$.

For the contribution $\chi_{10}^{(1)}$ to the (grand canonical) susceptibility one first calculates $\Delta F_{10}^{(1)}$, by performing the energy integral (2.54c), and then takes the field derivative twice. A leading-$\hbar$ calculation gives [45]

$$\frac{\chi_{10}^{(1)}}{\chi_L} = \frac{3(k_F a)^{3/2}}{8\pi^{1/2}} \frac{\sqrt{\cos\zeta}(\sin\zeta + \cos\zeta - 1)}{\zeta} \frac{[2\zeta - \sin(2\zeta)]^2}{\sin^4\zeta}$$

$$\times \sin\left(\frac{S_{10}}{\hbar} + \frac{\pi}{4}\right) R\left(\frac{L_{10}}{L_T}\right). \tag{4.86}$$

Figure 4.15a displays as the dashed line the entire bouncing-ball susceptibility $(\chi_{10}^{(1)} + \chi_{01}^{(1)})/\chi_L = 2\chi_{10}^{(1)}/\chi_L$ according to the above equality At fluxes up to $\varphi \approx 15$ it explains the low-frequency shift in the oscillations of the quantum result. Hence, the entire susceptibility at small fields is well approximated by $\chi_{11} + \chi_{10} + \chi_{01}$. The magnetic response is entirely governed by bouncing-ball periodic motion at fluxes between $\varphi \approx 15$ and $\varphi \approx 37$, the limit where $r_{cyc} = a/2$ and the last bouncing-ball orbits vanish. The agreement between the semiclassical and the full quantum result is excellent.

In order to illustrate that this agreement is not an artefact of the particular number of electrons chosen, we show in Fig. 4.15b semiclassical and quantum bouncing-ball oscillations for $k_B T/\Delta = 7$ and at a different Fermi energy corresponding to $\sim$1400 electrons. Upon decreasing the Fermi energy, the upper limit $r_{cyc} = a/2$ (or $k_F a/(2\pi\varphi) = 1/2$) of the bouncing-ball oscillations is shifted towards smaller fluxes ($\varphi \approx 30$ in Fig. 4.15b). Moreover, the number of corresponding oscillations decreases.

The flux dependence of the action S_{10} (see (4.81)), which is generally rather complicated, exhibits in the limit $a/r_{cyc} = 2\pi\varphi/(k_F a) \ll 1$ a quadratic dependence on φ [45]:

$$\frac{S_{10}}{\hbar} \simeq 2 k_F a \left[1 - \frac{1}{24}\left(\frac{2\pi\varphi}{k_F a}\right)^2\right]. \tag{4.87}$$

Using this expression for S_{10} in the susceptibility formula (4.86) gives the dotted curve in Fig. 4.15a. It gets out of phase with the full line at a flux corresponding to $a/r_{cyc} > 1$. While the period of the χ_{11} small-field oscillations is practically constant with respect to φ, we find a quadratic φ characteristic for the oscillations in the intermediate regime. In the strong-field regime, to be discussed in the next section, the oscillations exhibit a $1/\varphi$ behavior.

So far we have presented the magnetic response of the family of primitive orbits (1,0) and (0,1). It accurately describes the intermediate-field regime at rather high temperatures corresponding to a temperature cutoff length of the order of the system size. At low T one has to include higher repetitions $(j,0)$, $(0,j)$ along bouncing-ball orbits. The result of a corresponding calculation reads [45]:

$$\frac{\chi^{(1)}}{\chi_L} = \frac{1}{\chi_L} \sum_{j=1}^{\infty} \left(\chi_{j0}^{(1)} + \chi_{0j}^{(1)} \right)$$

$$= \frac{3}{4\pi^{1/2}} (k_F a)^{3/2} \frac{\sqrt{\cos\zeta}(\sin\zeta + \cos\zeta - 1)}{\zeta} \frac{[2\zeta - \sin(2\zeta)]^2}{\sin^4\zeta}$$

$$\times \sum_{j=1}^{\infty} j^{-1/2} \sin\left(j \frac{S_{10}}{\hbar} + \frac{\pi}{4} \right) R(j \, L_{10}/L_T) \, . \tag{4.88}$$

The susceptibility at the same Fermi energy as in Fig. 4.15b but at a significantly lower temperature $k_B T/\Delta = 2$ is displayed in Fig. 4.15c. The quantum mechanical bouncing-ball peaks are now much higher. Moreover, new peaks related to long periodic orbits differing from bouncing-ball trajectories appear. One finds that the bouncing-ball peak heights and even their shape, which is now asymmetric and no longer sinusoidal, is well reproduced by the analytical sum (4.88), showing the correct temperature characteristic of the semiclassical theory.

Bouncing-ball-type modulations should exist in general in microstructures in which parts of opposite boundaries are parallel and also, in particular, in circularly symmetrical microstructures such as a disk.

4.5.4 De Haas–van Alphen-Like Oscillations

Figure 4.1 shows that at strong field strengths or at small energy the spectrum of a square potential well exhibits the Landau fan corresponding to bulk-like Landau states almost unaffected by the system boundaries. In addtion, surface affected states fill the gaps between the Landau levels. They condense successively into the Landau channels with increasing magnetic field. These spectral features give rise to susceptibility oscillations which emerge with increasing amplitude for fluxes corresponding to $r_{cyc} < a/2$, for instance for $\varphi > 40$ in Fig. 4.2. These oscillations are shown in more detail in Fig. 4.16. The full line depicts the numerical quantum result. The susceptibility oscillations exhibit the same period $\sim 1/H$ as bulk de Haas–van Alphen oscillations but differ in amplitude, because here the cyclotron radius is not negligible compared to the system size.

In the extreme high-field regime $r_{cyc} \ll a$, where quantum mechanically the influence of the boundaries of the microstructure on the position of the quantum levels can be neglected (corresponding to the bulk case), the oscillations of the susceptibility are given by (4.14). Here, we summarize a semiclassical approach [45] which gives this equation and allows one moreover to include boundary effects.

In the high-field case only one type of primitive periodic orbit exists, namely the cyclotron orbits. Their length, enclosed area, and action are

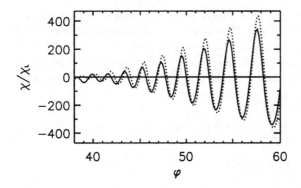

Fig. 4.16. High-field de Haas–van Alphen-like oscillations of the susceptibility of a square at magnetic fluxes corresponding to $r_{\mathrm{cyc}} < a/2$ for ~ 2100 electrons at $kT/\Delta = 8$. *Full line:* quantum calculations; *dotted line:* analytical semiclassical result from cyclotron orbits according to (4.93). (From [45], by permission)

$$L_{\mathrm{cyc}}(H) = 2\pi r_{\mathrm{cyc}} \ ,$$
$$A_{\mathrm{cyc}}(H) = -\pi r_{\mathrm{cyc}}^2 \ ,$$
$$\frac{S_{\mathrm{cyc}}(H)}{\hbar} = kL_{\mathrm{cyc}} + \frac{e}{c\hbar} HA_{\mathrm{cyc}} = k\pi r_{\mathrm{cyc}} \ . \tag{4.89}$$

Since the trajectory passes through a focal point after each half traversal along the cyclotron orbit, the Maslov indices are $\eta_n = 2n$ and the diagonal part of the semiclassical Green function reads, omitting the Weyl part of G,

$$G(\boldsymbol{r}, \boldsymbol{r}' = \boldsymbol{r}) = \frac{1}{\mathrm{i}\hbar\sqrt{2\pi\mathrm{i}\hbar}} \sum_n (-1)^n D_n \, \exp(\mathrm{i}n\pi k r_{\mathrm{cyc}}) \ . \tag{4.90}$$

A direct evaluation of the amplitude D_n in configuration space is difficult, since all trajectories starting at some point $\boldsymbol{r}$ refocus precisely at $\boldsymbol{r}$ (a focal point). Therefore, the usual representation (2.5) for the prefactors D_n is divergent and cannot be used. A method to overcome this problem by working with a Green function $\tilde{G}(x, y; p'_x, y')$ in the momentum representation for the x' direction instead of $G(x, y; x', y')$ is described in Appendix E of [45]. It yields

$$\frac{D_n}{\mathrm{i}\hbar\sqrt{2\pi\mathrm{i}\hbar}} = \frac{m}{\mathrm{i}\hbar^2} \ . \tag{4.91}$$

The oscillating part of the density of states is obtained after inserting the above expression into (4.90). It reads

$$d^{\mathrm{osc}}(E; H) = \sum_n d_n(E, H) = \frac{g_{\mathrm{s}} Am}{\pi\hbar^2} \sum_n (-1)^n \cos(n\pi k r_{\mathrm{cyc}}) \ . \tag{4.92}$$

From this relation one finds the de Haas–van Alphen susceptibility (4.14) for infinite systems upon taking the field derivatives. The contribution of cyclotron orbits to the susceptibility (4.14) has to be modified by the introduction of a multiplicative factor $s(H)$ when $r_{\rm cyc}$ is not negligible compared to a. This accounts for the fact that the family of periodic cyclotron orbits (not affected by the boundaries), which can be parametrized by the positions of the orbit centers, is diminished with decreasing field. One then obtains for a billiard-like quantum dot [45]

$$\frac{\chi_{\rm cyc}^{\rm GC}}{\chi_{\rm L}} = -6s(H)\,(k_{\rm F}r_{\rm cyc})^2 \sum_{n=1}^{\infty} (-1)^n R\left(\frac{2\pi n r_{\rm cyc}}{L_T}\right) \cos\left(n\pi k_{\rm F}r_{\rm cyc}\right) . \quad (4.93)$$

Here $s(H)$ is, for the case of the square,

$$s(H) = \left(1 - 2\frac{r_{\rm cyc}}{a}\right)^2 \Theta\left(1 - 2\frac{r_{\rm cyc}}{a}\right) , \quad (4.94)$$

where Θ is the Heaviside step function. Cyclotron orbits no longer fit into the square at a field corresponding to $r_{\rm cyc} = a/2$, i.e. $s(\varphi) = 0$. In Fig. 4.16 this is the case near $\varphi \approx 38$. The dotted line in the figure shows the semiclassical expression (4.93). It is in good agreement with the related numerical results. In particular, it accounts for the decrease in the amplitudes of the de Haas–van Alphen oscillations when approaching $\varphi(r_{\rm cyc} = a/2)$ from the strong-field limit. This behavior is specific to quantum dots. Corresponding de Haas–van Alphen oscillations in the two-dimensional bulk exhibit (nearly constant) amplitudes on the order of $\chi/\chi_{\rm L} \approx 3000$, when studied under the same conditions as for the curves in Fig. 4.16.

The energy-level diagram corresponding to the flux regime covered in Fig. 4.16 shows a complex variety of levels between the Landau manifolds (see Fig. 4.1). Hence it may appear surprising that the semiclassical curve in Fig. 4.16, reflecting only the contribution from unperturbed cyclotron orbits, agrees with the numerical curve, representing the complete system. Owing to finite temperature and the fact that angular momentum is not conserved in square billiards, the corresponding whispering-gallery or edge orbits are mostly chaotic and do not show up in the magnetic response. The strong de Haas–van Alphen-like oscillations may be considered as a manifestation of the dominant influence of the family of cyclotron orbits. We note that Sivan and Imry [271] observed additional high-frequency oscillations related to whispering-gallery orbits superimposed on the de Haas–van Alphen oscillations when studying the magnetization of a circular disk (where angular momentum is conserved) in the quantum Hall regime.

4.6 Systems Integrable at Arbitrary Fields: Ring Geometries

In Sect. 4.4 we addressed the generic situation where an applied magnetic field breaks the integrability of a system with regular dynamics at zero field:

the resulting orbital magnetic properties of such systems were derived and then illustrated for the case of the square in the preceding section.

There also exist "nongeneric" systems where the classical dynamics remains integrable in the presence of the magnetic field. A prominent example is the motion of a particle in a two-dimensional parabolic potential well in a uniform magnetic field. This potential is frequently used to model electronic properies of small quantum dots. The behavior of the energy levels as a function of the magnetic field was obtained analytically by Fock in 1928 [267] and its orbital magnetism was computed by Denton [233]. Other examples are disks and rings, geometries used in persistent current experiments, which belong, owing to their rotational symmetry, to the category of systems integrable at arbitrary field. The magnetic susceptibility of the circular billiard can be calculated from its exact wave mechanical solution in terms of Bessel functions [232, 235, 269]. The magnetic response of long cylinders [213, 268] and narrow rings [213], which represent the two nontrivial generalizations of one-dimensional rings, can be calculated by neglecting the curvature of the circle and solving the Schrödinger equation for a rectangle with periodic boundary conditions.

In these cases, and similarly for regular structures threaded by an Aharonov–Bohm flux, the Berry-Tabor trace formula (2.25) provides the appropriate means to calculate semiclassically the oscillating part of the density of states d^{osc}, including its field dependence. Then the free-energy terms $\Delta F^{(1)}$ and $\Delta F^{(2)}$, and their corresponding contributions to the magnetic response, can be deduced. Such a semiclassical calculation was first performed in [45, 69, 71] and will be summarized here. It provides an intuitive and unifying approach to the orbital magnetism of circular billiards and rings of any thickness, which allows one to establish the range of validity of previous studies. We add that an alternative approach to the magnetic susceptibility of various integrable geometries within the linear-response theory was presented in [276].

In [45] it was shown that the typical and average susceptibilities of circular billiards exhibit a large enhancement with respect to the bulk values by powers of $k_{\mathrm{F}}a$, which should allow for an experimental detection of these finite-size effects. However, to our knowledge there have not been any measurements of the magnetic response of electrons in disks so far. In the following we shall therefore place the corresponding findings for ring geometries in the foreground. They are of particular interest owing to recent experiments on the persistent current in regular ballistic rings [220].

4.6.1 Persistent Currents

Since we are dealing with integrable systems, the Berry–Tabor formula for the density of states ((2.25) introduced in Sect. 2.1.3), provides the natural starting point for a semiclassical treatment. The magnetic response at finite values of the perpendicular magnetic field can be obtained, at least in principle, from the calculation of the various classical entries in the Berry–Tabor formula at finite fields. This requires knowledge of the dependence of the classical dynamics on the magnetic field including, for example, the bending of classical orbits [275].

Here we again refer to the weak-field regime, where classical perturbation theory can be used to account for the modifications of the zero-field periodic trajectories. The most sensitive change is through the classical action and is given by $\delta S = (e/c)H\mathcal{A}$, where $\mathcal{A}$ is the algebraic area enclosed by the unperturbed orbit. In an integrable system the action is the same for all periodic orbits on a resonant torus. Hence, the fact that a system remains integrable at finite field implies that all the orbits of a family $\mathbf{M}$ enclose the same absolute area $\mathcal{A}_{\mathbf{M}}$, unlike, for example, to the case of the square. Using moreover the fact that the system is time-reversal invariant at zero field, and grouping pairs of time-reversal trajectories in (2.25), the density of states at weak fields reads

$$d_{\mathbf{M}}(E, H) = d_{\mathbf{M}}^0(E) \, \cos\left(\frac{2\pi\mathcal{A}_{\mathbf{M}}}{\phi_0}\right) . \qquad (4.95)$$

The field dependence is cosine-like, contrary to the generic integrable case, which is described by (4.48) and (4.51) in terms of a Bessel function and therefore decays with increasing field. This decay was related to the dephasing in the actions of orbits of the same family enclosing different areas.

The characteristics of the zero-field periodic orbits remain to be incorporated. For the case of two-dimensional ring billiards there are two types of periodic orbits as sketched in the insets of Fig. 4.17: orbits which do not touch the inner disk (type I) and those which hit it (type II).

Let us denote by a and b, respectively, the outer and inner radii of the ring. Type I and type II orbits are conveniently labeled by the topology $\mathbf{M} = (M_1, M_2)$, where M_1 is the number of circuits of the ring before returning to the initial point after M_2 bounces on the outer circle, and $M_2 \geq \hat{M}_2 = \text{Int}(M_1\pi/\arccos r)$ with $r = b/a$. The length and area of type I trajectories with topology $\mathbf{M}$ are

$$L_{\mathbf{M}} = 2M_2 a \, \sin\delta \quad , \quad \mathcal{A}_{\mathbf{M}} = \frac{M_2 a^2}{2} \sin 2\delta , \qquad (4.96)$$

where $\delta = \pi M_1/M_2$. For type II trajectories one finds [45]

$$\tilde{L}_{\mathbf{M}} = 2M_2 a \sqrt{1 + r^2 - 2r\cos\delta} \quad , \quad \tilde{A}_{\mathbf{M}} = M_2 ab \sin\delta . \qquad (4.97)$$

Following Keller and Rubinow [270], the action integrals $\boldsymbol{I} = (I_1, I_2)$ and the function g_E which enter into the Berry–Tabor trace formula (2.25) were calculated in [45]. The contributions to the zero-field density of states from trajectories of type I and type II and topology $\mathbf{M}$ then read

$$d_{\mathbf{M}}^0(E) = \sqrt{\frac{2}{\pi}} \frac{g_{\mathrm{s}} m L_{\mathbf{M}}^{3/2}}{\hbar^2 k^{1/2} M_2^2} \cos\left(k L_{\mathbf{M}} + \frac{\pi}{4} - \frac{3\pi}{2} M_2\right), \qquad (4.98a)$$

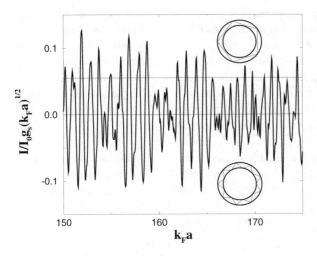

Fig. 4.17. Persistent current in a single ring ($r = b/a = 0.9$) for a field corresponding to a flux of $\phi_0/4$. The typical current is shown by the *upper horizontal line. Upper* and *lower insets*: type I and type II trajectories (from [69], by permission)

$$\tilde{d}_M^0(E) = 4\sqrt{\frac{2}{\pi}} \frac{g_s a^2 m}{\hbar^2 (k\tilde{L}_M)^{1/2}} \sin\left(k\tilde{L}_M + \frac{\pi}{4}\right)$$

$$\times \left[(1 - r\cos\delta)(r\cos\delta - r^2)\right]^{1/2} . \tag{4.98b}$$

The persistent current in a ring was defined in (4.16) as $I = -c(\partial F/\partial\phi)_{T,N}$. We use the area $A = \pi a^2$ of the outer circle as the defining area: $\phi = AH$. Applying (2.54c) and (4.23b), the persistent current of a ring billiard can be expressed as the sum of two contributions corresponding to both types of trajectories [69]:

$$\frac{I^{(1)}}{I_0} = g_s (k_F a)^{1/2} \sum_{M_1, M_2 \geq \hat{M}_2} \left[\mathcal{I}_{M,I}^{(1)} \sin\left(\frac{eH}{\hbar c} A_M\right) R\left(\frac{L_M}{L_T}\right) \right.$$

$$\left. + \mathcal{I}_{M,II}^{(1)} \sin\left(\frac{eH}{\hbar c} \tilde{A}_M\right) R\left(\frac{\tilde{L}_M}{L_T}\right) \right] , \tag{4.99}$$

with

$$\mathcal{I}_{M,I}^{(1)} = 2\sqrt{\frac{2}{\pi}} \frac{1}{M_2^2} \frac{(A_M/a^2)}{(L_M/a)^{1/2}} \cos\left(k_F L_M + \frac{\pi}{4} - \frac{3\pi}{2} M_2\right) , \tag{4.100a}$$

$$\mathcal{I}_{M,II}^{(1)} = \sqrt{\frac{2}{\pi}} \frac{(\tilde{A}_M/a^2)}{(\tilde{L}_M/a)^{5/2}} \left[(1 - r\cos\delta)(r\cos\delta - r^2)\right]^{1/2}$$

$$\times \sin\left(k\tilde{L}_M + \frac{\pi}{4}\right) \tag{4.100b}$$

where $I_0 = ev_F/2\pi a$.

In Fig. 4.17 we show the persistent current according to (4.99) for a thermal length $L_T = 10a$ and a magnetic field $H = \phi_0/(4A)$. We are thus focusing on the first harmonic of I, and the whole sum can be obtained by keeping only the $M_1 = 1$ terms. Even for other fields, the higher winding numbers ($M_1 > 1$) are strongly suppressed at the temperatures of experimental relevance because L_M is roughly proportional to M_1. The persistent current of a given sample can be paramagnetic or diamagnetic, with a characteristic period in k given by the circle perimeter $L = 2\pi a$.

In order to characterize the typical value $I^{(t)} = [\overline{(I^{(1)})^2}]^{1/2}$ of the magnetic response we average I^2 over a $k_F a$ interval containing many oscillations, but negligible on the classical scale. This gives the upper horizontal line of Fig. 4.17. At finite temperatures and for a sufficiently large integration interval Δk, the nondiagonal terms of I^2 involving two different families of trajectories are unimportant and one gets [69]

$$\frac{I^{(t)}}{I_0} \simeq g_s (k_F a)^{1/2} \sum_{M_1, M_2 \geq \hat{M}_2} \left[\left(\mathcal{I}_{M,I}^{(t)} \right)^2 \sin^2 \left(\frac{eH}{\hbar c} \mathcal{A}_M \right) R^2 (L_M/L_T) \right.$$

$$\left. + \left(\mathcal{I}_{M,II}^{(t)} \right)^2 \sin^2 \left(\frac{eH}{\hbar c} \tilde{\mathcal{A}}_M \right) R^2 (\tilde{L}_M/L_T) \right]^{1/2}, \qquad (4.101)$$

where $(\mathcal{I}_{M,I}^{(t)})^2$ and $(\mathcal{I}_{M,II}^{(t)})^2$ are obtained from (4.100a) and (4.100b) simply by replacing the average of $\cos^2 (k_F L_M + \pi/4 - 3 M_2 \pi/2)$ and $\sin^2 (k_F \tilde{L}_M + \pi/4)$ by $1/2$.

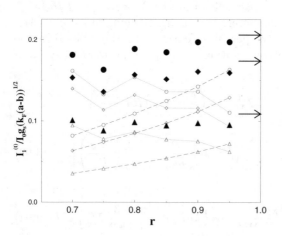

Fig. 4.18. Typical persistent current in rings of different thickness ($r = b/a$) for cutoff lengths $L_T = 30a$ (*circles*), $6a$ (*diamonds*), and $3a$ (*triangles*). *Filled symbols* denote the total current and lie approximately on a straight line approaching the asymptotic limit (see (4.103), indicated by *arrows*). *Unfilled symbols* represent the contributions from both types of trajectories and are joined by *dotted* (type I) and *dashed* (type II) guide-to-the-eye lines. (From [69], by permission)

In Fig. 4.18 we present the typical persistent current and its two contributions for various ratios $r = b/a$ and cutoff lengths L_T for the first harmonic ($M_1 = 1$). The contribution $\mathcal{I}_{M,\mathrm{I}}^{(\mathrm{t})}$ of type I trajectories dominates for small r (where the inner circle is not important and we recover the magnetic response of the circular billiard), while type II trajectories take over for narrow rings. The crossover value of r depends on temperature through L_T owing to the different dependences of the trajectory length on M (see (4.96) and (4.97)) for the two types of trajectories.

Here we have presented results for the typical current; at present these are of more relevance than the related results for the average current, which can be found in [45].

Thin Rings

Thin rings ($a \simeq b, r \simeq 1$) are particularly interesting since they are the configuration used in the experiment of [220], and further approximations can be performed on (4.100a) and (4.100b) using $(1 - r)$ as a small parameter. This gives a more compact and meaningful expression for the typical persistent current.

Since $\hat{M}_2 \simeq \pi/[2(1 - r)]^{1/2} \gg 1$, for $M_2 \geq \hat{M}_2$ the length and area of contributing orbits can be approximated by $L_M \simeq L = 2\pi a$ and $\mathcal{A}_M \simeq \tilde{A}_M \simeq A = \pi a^2$. For type II trajectories we have $\tilde{L}_M \simeq L$ for $M_2 \simeq \hat{M}_2$, and $\tilde{L}_M \simeq 2M_2(a - b)$ when $M_2 \gg \hat{M}_2$. Even if the main contributions $(\mathcal{I}_{M,\mathrm{II}})^2$ come from $M_2 \simeq \pi/[5^{1/6}(1-r)^{2/3}] \gg \hat{M}_2$, their associated $\tilde{L}_M$ are still of the order of L (to leading order in $1 - r$). One finds [69]

$$\frac{I^{(\mathrm{t})}}{I_0} = g_{\mathrm{s}}(k_{\mathrm{F}}a)^{1/2} \left(\sum_{M_1, M_2 \geq \hat{M}_2} \left\{ \frac{M_1}{M_2^4} + \frac{M_1^2 \pi (1-r)^2}{M_2^5 [(1 - r)^2 + \delta^2]^{5/2}} \right\} \right.$$

$$\left. \times \sin^2 \left(\frac{2\pi M_1 \phi}{\phi_0} \right) R^2(M_1 L/L_T) \right]^{1/2} . \qquad (4.102)$$

Since $\hat{M}_2 \gg 1$ we can convert the previous sums into integrals. To leading order in $1 - r$ the persistent current is dominated by type II trajectories and is of the form

$$\frac{I^{(\mathrm{t})}}{I_0} = \frac{2g_{\mathrm{s}}}{\pi\sqrt{3}} \sqrt{\mathcal{N}} \left[\sum_{M_1} \frac{1}{M_1^2} \sin^2 \left(2\pi M_1 \frac{\phi}{\phi_0} \right) R^2 \left(\frac{M_1 L}{L_T} \right) \right]^{1/2} , \qquad (4.103)$$

consistent with the result of [213].

Comparison with Experiment

In the experiment by Mailly, Chapelier, and Benoit [220] persistent currents were measured in a single thin semiconductor ring (with effective outer and inner radii $a = 1.43$ μm and $b = 1.27$ μm). The experiment was in the ballistic and phase-coherent regime ($l = 11$ μm and $L_\phi = 25$ μm). The number of occupied channels was $\mathcal{N} \simeq 4$. The quoted temperature of $T = 15$ mK makes the temperature factor irrelevant for the first harmonic ($L_T \simeq 30a$, $R(L/L_T) \simeq 1$). The magnetic response exhibits an hc/e flux periodicity and changes from diamagnetic to paramagnetic on changing the microscopic configuration, consistent with (4.99)–(4.100b). Unfortunately, the sensitivity was not high enough to be able to test the signal averaging with these microscopic changes. The typical persistent current was found to be 4 nA, while (4.103) and [213] would yield 7 nA. However, the difference between the theoretical and measured values is not significant, given the experimental uncertainties as discussed in [220].

As in the case of the susceptibility of squares, residual disorder effects (see Chap. 5) and interaction effects (see Chap. 6) must be considered for a more detailed comparison with the experiment. Nevertheless, new experiments on individual rings and on ensembles of ballistic rings are desirable as a test of the semiclassical approach of this section.

5. Disorder Effects

Part of the sustained interest during the last decade in the physics of two-dimensional electron gases and mesoscopic systems stems from the high electron mobilities that can be achieved at low temperatures. The limiting factor for the mobility in this regime, namely impurity scattering, can be strongly reduced in state-of-the-art semiconductor heterojunctions. These technological achievements have motivated theoretical approaches where the actual microstructure is approximated by *clean* quantum billiards ignoring impurity scattering completely.

Within these simple models of a particle-in-a-billiard, important differences have been predicted for instance in the transport through chaotic and integrable cavities, as well as for their orbital magnetism as discussed in the previous chapters.

When addressing the issue of observable signatures of chaotic and integrable dynamics for actual microstructures, the role of residual impurities has to be considered since small amounts of disorder are unavoidable. It has been shown that even weak disorder can be strong enough to mix energy levels, influence spectral statistics [278, 279], and affect related thermodynamic quantities [283]. Indeed, any perturbing potential, such as the one provided by the disorder, immediately breaks the integrable character of the classical dynamics. Therefore, the question of whether or not integrable behavior should be observed naturally arises. It is hence of interest to establish whether the differences between chaotic and integrable clean geometries persist beyond the particle-in-a-box model. Therefore we shall address in this section mesoscopic effects arising from the interplay between confinement and disorder potentials.

5.1 Characterization

We begin with a characterization of the disorder. The two opposite limits are the clean case, where the dynamics is completely governed by the nonrandom confinement potential, and the *diffusive* regime, where the electron motion equals a random walk between the impurities. In the latter, confining effects are not important, at least on timescales smaller than the Thouless time, the time a particle needs to diffuse across the system. The disorder in the diffusive case is characterized by two length scales. The *elastic mean free path* (MFP) l is associated with the total amplitude diffracted by the disorder [280]. It is related to the single-particle relaxation time [281]. The *transport mean free path* l_T denotes the mean distance over which the electron momentum is randomized. In a more formal approach, l is given by the impurity average of the one-particle Green function, while l_T is related to the two-particle Green function. Hence, l_T has a classical meaning, while l is a quantum mechanical quantity.

The condition $l_T \ll a$ defines the diffusive limit. When l_T is of the order of the typical size a of the microstructure, confinement *and* disorder are relevant. The *ballistic* regime is defined for $l_T > a$, where electrons traverse the structure with a small drift in their momentum traveling along almost straight lines. Then their dynamics is mainly given by reflections at the walls of the confining potential and the underlying classical mechanics still depends on the geometry. In this regime we shall investigate the different roles of disorder in integrable and chaotic geometries.

For short-range impurity potentials, as are typically found in metallic samples, the scattering is isotropic and the momentum is randomized after each collision with an impurity. Hence, $l_T \simeq l$ and there is only one length scale characterizing the disorder. In high-mobility semiconductor heterojunctions the modulation-doping technique allows the spatial separation of the impurities and the conducting electrons, yielding a relatively smooth random potential at the level of the two-dimensional electron gas (2DEG). As will be shown, smooth-disorder effects, depending on the ratio between the finite system size a and the disorder correlation length ξ, are specific to confined systems which exhibit strong deviations from the bulk behavior. For smooth impurity potentials the scattering is forward-directed and l_T may be significantly larger than l [280]. The regime $l_T > a > l$ is particularly interesting: it is ballistic because the classical mechanics is hardly affected by disorder, but the single-particle eigenstates are short-lived. Therefore both the confinement *and* the impurities have to be considered.

The general purpose of this chapter is to study weak-disorder effects in constrained geometries with noninteracting particles. This is a fundamental problem since, contrary to the bulk, a disorder-averaged confined system is not translationally invariant and one has to impose in quantum calculations the correct boundary conditions of the geometry. The conventional techniques used to deal with impurity scattering such as diagrammatic perturbation theory efficiently describe the diffusive regime [105], but calculations become rather involved when the confinement and the detailed nature of the impurity potential have to be considered. Supersymmetry and random-matrix theories have been widely used in recent studies of quantum chaos and disordered systems [198–200]; however, they are not directly applicable to our regime of interest, which includes the short-time behavior and disorder effects in regular geometries. There, disorder effects are expected to be most significant, because the weak random potential perturbs the integrable dynamics.

In defining spectral correlation functions in the ballistic regime or, more generally, for finite disordered systems, one has to distinguish between disorder averaging, which we denote by $\langle \ldots \rangle$, and size or energy averaging which we denote by an overbar. Pure disorder averaging corresponds to the experimental situation of an ensemble of weakly disordered microstructures of the same size. Recent work has shown that for this case spectral correlation functions contain, and often are dominated by structures that are strongly oscillationg in energy on the scale of $k_F a$, reflecting the presence of the confinement [278, 283].

The consideration of spectral correlations after both energy and (independent) disorder averaging corresponds to the experimental situation of an ensemble of disordered systems with variation in their sizes. Then it is convenient to divide the two-level correlation function, $K(\varepsilon_1, \varepsilon_2)$ (2.28), into two separate terms [73, 278]:

$$K(\varepsilon_1, \varepsilon_2; H) \equiv \overline{K^d(\varepsilon_1, \varepsilon_2; H)} + K^s(\varepsilon_1, \varepsilon_2; H). \tag{5.1}$$

Here

$$K^d(\varepsilon_1, \varepsilon_2) = \left[\langle d(\varepsilon_1) d(\varepsilon_2) \rangle_d - \langle d(\varepsilon_1) \rangle \langle d(\varepsilon_2) \rangle \right] / \bar{d}^2 , \tag{5.2a}$$

$$K^s(\varepsilon_1, \varepsilon_2) = \overline{\langle d(\varepsilon_1) \rangle \langle d(\varepsilon_2) \rangle} / \bar{d}^2 - 1 , \tag{5.2b}$$

where d denotes the single particle density of states and $\bar{d}$ its mean part. K^d is a measure of $disorder\text{-}induced$ correlations of the density of states, while K^s is given by $size\text{-}induced$ correlations. In a diffusive system, the disorder-averaged density of states $\langle d(\varepsilon) \rangle$ is a constant $\simeq \bar{d}$, so that K^s is vanishingly small and K^d dominates. However, for $l > a$ both correlation functions may be relevant and, once $l > a(k_F a)^{d-1}$, disorder-induced mixing of levels is negligible and K^s prevails.

In this chapter we review a general treatment of disorder effects in nondiffusive confined systems employing semiclassical Green functions. In a semiclassical picture two types of paths can be distinguished in such systems:

(i) Trajectories which exist also in the absence of disorder and hence contribute to the Green function of the clean system. They yield size-induced contributions to K^s which are damped upon impurity scattering.

(ii) Paths which explicitly result from impurity scattering. They contribute to K^d and vanish in the clean limit.

In the ballistic case at finite temperature the former are of major importance. On the basis of those paths, we study ballistic disorder effects semiclassically in the next subsection. In Sects. 5.3 and 5.4 we illustrate the results obtained by considering impurity effects on that part of the orbital magnetism which is sensitive to size-induced spectral correlations as measured by $K^s(\varepsilon_1, \varepsilon_2)$.

In Sect. 5.5 we then consider disorder-induced correlations K^{d} from paths of type (ii) and extend the semiclassical approach based on paths (i) to describe the complete crossover from the clean limit to the opposite diffusive limit in constrained systems [73, 282].

5.2 Semiclassical Treatment of Disorder in the Ballistic Limit

Disorder has usually been studied for macroscopic metallic samples which are self-averaging or for ensembles of mesoscopic samples where different structures present different impurity configurations. The possibility of measuring a single disordered mesoscopic device poses a conceptual difficulty since there is no averaging process directly involved. When discussing disorder effects on the orbital magnetism of microstructures, it is therefore necessary to distinguish between the behavior of an individual sample and an ensemble [278]. Moreover, we have to distinguish the cases where the Fermi energy and the size of the microstructures are kept fixed under impurity averaging from the cases where these parameters change with the different impurity realizations.

These various averages, which will be discussed in the following, can be expressed in terms of the impurity averages of one- and two-particle Green functions. Following [72, 283], we begin in this section with a general treatment of disorder effects in terms of semiclassical expansions of Green functions (based on paths of type (i)). A similar semiclassical approach to smooth disorder, but restricted to the bulk case, was developed in [284]. As an application of this approach we study in Sect. 5.3 the case of a 2DEG in a magnetic field in the presence of disorder and calculate disorder effects on the susceptibility of quantum dots (Sect. 5.4.2). In particular, we demonstrate that for integrable structures the effect of smooth disorder results in a power-law damping instead of an exponential behavior.

5.2.1 The Disorder Model

We do not intend to calculate a disorder potential from realistic distributions of residual impurities in semiconductor heterostructures on the basis of microscopic models (e.g. [285]). Instead, we assume in the following that the disorder potential $V^{\mathrm{dis}}(\boldsymbol{r})$ is spatially random and characterized by the correlation function

$$C(|\boldsymbol{r} - \boldsymbol{r}'|) = \langle V^{\mathrm{dis}}(\boldsymbol{r}) V^{\mathrm{dis}}(\boldsymbol{r}') \rangle . \tag{5.3}$$

The mean disorder strength is $C^0 \equiv C(0)$ and the typical correlation length is denoted by ξ. In order to allow for the derivation of analytical expressions for disorder averages we shall use a Gaussian correlation function

$$C(|\boldsymbol{r} - \boldsymbol{r}'|) = C^0 \exp\left(-\frac{(\boldsymbol{r} - \boldsymbol{r}')^2}{4\xi^2}\right) . \tag{5.4}$$

Disorder effects in the ballistic regime depend on different length scales: the disorder correlation length ξ, the Fermi wavelength λ_F of the electrons, and the system size a. In the bulk case of an unconstrained 2DEG we shall distinguish between *short-range* ($\xi < \lambda_F$) and *finite-range* ($\xi > \lambda_F$) disorder potentials. In the case of a microstructure a third, *long-range* regime for $\xi > a > \lambda_F$ has to be considered.

The disorder correlation function (5.4) may be regarded as being generated by a realization i of a two-dimensional Gaussian disorder potential given by the sum

$$V^{\mathrm{dis}}(\boldsymbol{r}) = \sum_{j}^{N_i} \frac{u_j}{2\pi\xi^2} \exp\left(-\frac{(\boldsymbol{r} - \boldsymbol{R}_j)^2}{2\xi^2}\right) \tag{5.5}$$

of the potentials of N_i *independent* impurities j. They are located at points $\boldsymbol{R}_j$ and distributed uniformally over the area A. The strengths u_j in the above equality are uncorrelated: $\langle u_j u_{j'}\rangle = u^2\delta_{jj'}$. For this model, the disorder strength (5.4) is

$$C^0 = \frac{u^2 n_i}{4\pi\xi^2} \tag{5.6}$$

with $n_i = N_i/A$. In Appendix A.3 quantum mechanical expressions for the transport MFP l_T and the elastic MFP l in terms of the parameters n_i and u_i are provided. The white-noise case of δ-function scatterers, $V^{\mathrm{dis}}(\boldsymbol{r}) = \sum_j^{N_i} u_j \delta(\boldsymbol{r} - \boldsymbol{R}_j)$, is reached in the limit $\xi \to 0$.

The above model with a finite ξ is considered to appropriately account for the smooth-disorder potentials existing in heterostructures. For the cleanest samples used in today's experiments the characteristic scale ξ can be on the order of 100–200 nm [291]. Thus, these systems typically operate in the finite-range regime $a > \xi > \lambda_F$.

The Gaussian disorder model will be helpful for some of the analytical calculations and is used for numerical quantum simulations (see Appendix A.2). Nevertheless, general results based on the correlation function $C(|\boldsymbol{r} - \boldsymbol{r}'|)$ in (5.4) do not rely on the particular choice of this disorder potential.

5.2.2 Effect on the Single-Particle Green Function

We begin with the investigation of disorder effects on the single-particle Green function, given in the semiclassical representation (2.3) as a sum over classical paths in a clean system. In the finite-range or long-range regime, where the disorder potential is smooth on the scale of λ_F, a semiclassical treatment is well justified (given a microstructure with size $a \gg \lambda_F$).

The classical mechanics of paths with length $L_t \ll l_T$ is essentially unaffected by disorder. The trajectories themselves as well as their classical amplitudes D_t and topological indices η_t can be considered as unchanged. In the ballistic limit the dominant disorder effect on the Green function in (2.3) comes from shifts in the semiclassical phases due to the modification of the classical actions. According to the semiclassical perturbative approach in Sect. 2.1 the classical action along a path $\mathcal{C}_t$ in a system with a weak disorder potential ($V^{\mathrm{dis}} \ll E$) can be written as (see (2.12) and (2.13))

$$S_t^{\mathrm{d}} \simeq S_t + \delta S_t \quad ; \quad \delta S_t = -\frac{1}{v_{\mathrm{F}}} \int_{\mathcal{C}_t} V^{\mathrm{dis}}(\boldsymbol{q}) \, dq \, . \tag{5.7}$$

The action is given to leading order by the action S_t along the *unperturbed* trajectory $\mathcal{C}_t$ of the clean system. The disorder enters via the correction term δS_t. In this approximation a disorder average $\langle \ldots \rangle$ acts only on this term. The impurity-averaged Green function then reads

$$\langle G(\boldsymbol{r}, \boldsymbol{r}'; E) \rangle = \sum_t G_t(\boldsymbol{r}, \boldsymbol{r}'; E) \, \langle \exp\left[(\mathrm{i}/\hbar)\delta S_t\right] \rangle \, . \tag{5.8}$$

Here G_t denotes the contribution of the path t to the Green function without disorder.

For orbits of length $L_t \gg \xi$ disorder contributions to δS (according to (5.7)) at trajectory segments which are separated by a distance larger than ξ are uncorrelated. The accumulation of action along the path is therefore stochastic and can be interpreted as being determined by a random-walk process. This results in a Gaussian distribution of $\delta S_t(L_t)$. For larger ξ ($L_t \gg \xi$), the dephasing δS_t can still be considered to follow a Gaussian distribution law provided the disorder potential is generated by a sum of a large number of independent impurity potentials. Given the Gaussian character of the $\delta S_t(L_t)$ distribution, the disorder term in (5.8) can be written as

$$\langle \exp\left[(\mathrm{i}/\hbar)\delta S_t\right] \rangle = \exp\left[-\langle \delta S_t^2 \rangle/(2\hbar^2)\right] \, . \tag{5.9}$$

It is completely specified by the Gaussian variance

$$\langle \delta S_t^2 \rangle = \frac{1}{v_{\mathrm{F}}^2} \int_{\mathcal{C}_t} dq \int_{\mathcal{C}_t} dq' \langle V^{\mathrm{dis}}(\boldsymbol{q}) V^{\mathrm{dis}}(\boldsymbol{q}') \rangle \, , \tag{5.10}$$

which is the mean of the disorder correlation function $C(|\boldsymbol{q} - \boldsymbol{q}'|)$ along the unperturbed trajectory.

Bulk Case

We begin with the evaluation of the variance for the two-dimensional bulk contribution to the Green function in (5.8). In the bulk case there is only a single straight-line trajectory from $\boldsymbol{r}'$ to $\boldsymbol{r}$. If $L = |\boldsymbol{r}' - \boldsymbol{r}| \gg \xi$ the limits of the inner integral in (5.10) can be extended to infinity and we obtain for the variance

$$\langle \delta S^2 \rangle = \frac{L}{v_{\mathrm{F}}^2} \int dq \, C(\boldsymbol{q}) \, . \tag{5.11}$$

In view of (5.9), the semiclassical average Green function for an unconstrained system exhibits an exponential behavior [72, 283, 284]

$$\langle G(\boldsymbol{r}', \boldsymbol{r}; E) \rangle = G(\boldsymbol{r}', \boldsymbol{r}; E) \exp(-L/2l) \qquad (5.12)$$

on length scales $l_{\mathrm{T}} > L \gg \xi$. The damping exponent can be associated with an inverse *elastic* mean free path

$$\frac{1}{l} = \frac{1}{\hbar^2 v_{\mathrm{F}}^2} \int \mathrm{d}q \, C(\boldsymbol{q}). \qquad (5.13)$$

For a Gaussian correlation $C(\boldsymbol{q})$ of the form given in (5.4) the elastic MFP reads

$$l = \frac{\hbar^2 v_{\mathrm{F}}^2}{\xi \sqrt{\pi} C^0}. \qquad (5.14)$$

Using the mean disorder strength (5.6) for the potential of random Gaussian impurities, we get

$$l = \frac{4\sqrt{\pi}\hbar^2 v_{\mathrm{F}}^2 \xi}{u^2 n_i}, \qquad (5.15)$$

which relates l directly to the parameters n_i, u, and ξ of the impurity potential.

As discussed in Appendix A.3, the semiclassical result (5.15) agrees asymptotically to leading order in $k_{\mathrm{F}}\xi$ with the corresponding result ((A.46) in Appendix A.3) for the elastic MFP obtained from quantum diagrammatic perturbation theory for the bulk.

For small ξ, especially for $\xi < \lambda_{\mathrm{F}}$, the semiclassical approach reviewed here is no longer applicable. For white noise the self-consistent Born approximation leads to an integral equation for the disorder-averaged single-particle Green function [286]:

$$\langle G^+(\boldsymbol{r}, \boldsymbol{r}') \rangle = G^+(\boldsymbol{r}, \boldsymbol{r}') \qquad (5.16)$$

$$+ \frac{\hbar}{2\pi\tau(\bar{d}/A)} \int \mathrm{d}\boldsymbol{r}'' \, G^+(\boldsymbol{r}, \boldsymbol{r}'')\langle G^+(\boldsymbol{r}'', \boldsymbol{r}'')\rangle\langle G^+(\boldsymbol{r}'', \boldsymbol{r}')\rangle.$$

In the bulk case $\langle G^+(\boldsymbol{r}'', \boldsymbol{r}'')\rangle$ is given semiclassically in terms of "paths of zero length". The solution of the integral equation then leads to the exponential disorder damping of the (free) Green function. Hence in this regime (5.12) still holds, but with $l = v_{\mathrm{F}}\tau$ replaced by l_δ given in (A.45) of Appendix A.3. This damping can also be related to the (constant) self-energy occurring in the averaged Green function (2.66) within the Born approximation.

Confined System

For confined systems further paths of finite length can in principle contribute to $\langle G^+(\boldsymbol{r}'', \boldsymbol{r}'')\rangle$ in (5.16). They can be semiclassically viewed as paths scattered off the δ-like impurities, which may lead to additional corrections to $\langle G^+(\boldsymbol{r}, \boldsymbol{r}')\rangle$. These contributions are, however, of higher order in $\hbar$ and $1/\tau$.

We now consider (smooth) disorder effects in the presence of a confinement potential in the ballistic regime where the tranport MFP l_T is much larger than the system size a. In Appendix A.3 it is shown that for finite ξ the transport MFP is considerably larger than the elastic one. Thus, a ballistic treatment is well justified, even if l is of the order of the system size.

Confinement implies that the semiclassical Green function $G(\mathbf{r}, \mathbf{r}'; E)$ is given as a sum over all direct *and* multiply reflected paths connecting $\mathbf{r}'$ and $\mathbf{r}$; the disorder potential modifies the corresponding actions according to (5.7).

For short- and finite-range scatterers, (5.11) can be used, and therefore the damping of each path contribution $\langle G_t(E) \rangle$ to $\langle G(E) \rangle$ is given by

$$\langle G(\mathbf{r}, \mathbf{r}'; E) \rangle = \sum_t G_t(\mathbf{r}, \mathbf{r}'; E) \ \exp\left(-\frac{L_t}{2l}\right) . \tag{5.17}$$

These expression is analogous to the bulk case (5.12), but L is now replaced by the trajectory length $L_t > a \gg \xi$. We find an individual damping $\exp(-L_t/2l)$ for each geometry-affected orbit contributing to $\langle G_E \rangle$.

In the long-range case for $\xi \geq a$ correlations across different segments of an orbit become important. Thus, the correlation integral (5.10) can no longer be approximated by $L \int_{-\infty}^{+\infty} dq \, C(q)$ (as for $\xi \ll L_t$). Hence, the orbit geometry enters into the correlation integral and it generally cannot be evaluated analytically. However, one can expand $C(|\mathbf{r} - \mathbf{r}'|)$ for $\xi \gg a$ and obtain for Gaussian disorder, to first order in ξ^{-2},

$$C(|\mathbf{r} - \mathbf{r}'| \simeq C^0 \left[1 - \frac{(\mathbf{r} - \mathbf{r}')^2}{4\xi^2} \right] . \tag{5.18}$$

Using this approximation in the integral (5.10) gives an exponent

$$\frac{\langle \delta S_t^2 \rangle}{2\hbar^2} = \frac{1}{4\sqrt{\pi}} \frac{L_t^2}{l\xi} \left(1 - \frac{1}{2} \frac{I_t}{\xi^2} \right) , \tag{5.19}$$

which governs the damping of the Green function. In the above equation $I_t = (1/L_t) \int_{\mathcal{C}_t} r^2(q) \, dq$ can be regarded as the "moment of inertia" of the unperturbed trajectory $\mathcal{C}_t$ with respect to its "center of mass" $(1/L_t) \int_{\mathcal{C}_t} r(q) \, dq$. The damping in the long-range regime has specific features: according to (5.19) the damping depends quadratically on L_t, contrary to the linear behavior in the finite-range and bulk cases. The length scale of the damping is now given by the geometric mean of the bulk MFP l *and* ξ. The leading damping term does not depend on the specific orbit geometry, since it essentially reflects fluctuations in the mean of the smooth potentials of different impurity configurations. Inclusion of higher powers of ξ^{-2} leads to additional damping contributions from higher moments $\int_{\mathcal{C}_t} r^n(q) \, dq$.

The implications of the disorder average for the semiclassical density of states are obvious. For instance, contributions from primitive periodic orbits (ppo) in semiclassical trace formulas are exponentially damped according to

(5.12) and (5.19) for the finite- and long-range regimes, respectively. Higher repetitions j of a primitive periodic orbit exhibit a damping exponent $\sim j^2$, since the path experiences j times the same disorder configuration. This behavior is most easily seen from (5.9), with $\delta S_t = j \delta S_{\mathrm{ppo}}$. It plays a role in the line shape of impurity-broadened Landau levels, which will be discussed in Sect. 5.3.

The disorder damping of semiclassical single-particle Green functions has already been used in the semiclassical trace formulas for the conductivity (Chap. 3). There the quadratic damping ($\sim j^2$) was neglected, since higher repetitions were of minor importance owing to the additional temperature cutoff. Further spectral implications of the disorder-damped single-particle Green functions will be considered in Sect. 5.4.1.

5.2.3 Effect on the Two-Particle Green Function

Density correlation functions in general (see Sect. 2.2 and (5.1)) and spectral quantities such as the typical susceptibility (4.24a) and the ensemble-averaged susceptibility (4.24b) involve squares of the density of states. The latter can be expressed in terms of the difference between retarded and advanced Green functions $(G^+ - G^-)$, giving rise to products of one-particle Green functions. The cross products $G^+(\boldsymbol{r}, \boldsymbol{r}')\, G^-(\boldsymbol{r}, \boldsymbol{r}') = G^+(\boldsymbol{r}, \boldsymbol{r}')G^{+*}(\boldsymbol{r}', \boldsymbol{r})$ are of special interest, because they survive the energy averaging and are sensitive to changes in the magnetic field. Because the general two-particle Green function factorizes into a product of one-particle Green functions [287] in the approach of noninteracting particles used in this section, we shall use the former as a synonym for the latter.

We begin with a discussion of the underlying ideas of the semiclassical average of products of single-particle Green functions. This will be quantitatively evaluated for the susceptibility of microstructures in Sect. 5.4.2.

Let us consider a typical product $G(\boldsymbol{r}, \boldsymbol{r}')G^*(\boldsymbol{r}', \boldsymbol{r})$. As for the ballistic single-particle case, the effect of the disorder potential on each of the Green functions can be approximated by using (5.7). Then the disorder-averaged two-particle Green function, which contributes to K^{s} (see (5.2b)),[1] is given as a double sum over the averaged contributions from paths t and t':

$$
\begin{aligned}
\langle G(E)G(E)^* \rangle &= \sum_t \sum_{t'} \langle G_t(E)\, G_{t'}^*(E) \rangle \\
&= \sum_t \sum_{t'} G_t(E)\, G_{t'}^*(E) \langle \mathrm{e}^{(\mathrm{i}/\hbar)(\delta S_t - \delta S_{t'})} \rangle \\
&= \sum_t \sum_{t'} G_t(E)\, G_{t'}^*(E) \exp\left[-\frac{\langle (\delta S_t - \delta S_{t'})^2 \rangle}{2\hbar^2} \right] .
\end{aligned}
\tag{5.20}
$$

---

[1] Corresponding contributions to K^{d} will be discussed in Sect. 5.5.

In the third equality we have made use of (5.9). For the evaluation of the exponent it is now necessary to take into account disorder correlations between points on trajectories t and t'.

If $t=t'$ or the path is equal to its time reversal, the orbits acquire the same phase shift and therefore $\langle G_t(E)\, G_t^*(E)\rangle = |G_t(E)|^2$: within the approximation that disorder enters only in the phases, the diagonal contributions $t=t'$ remain the same as for clean systems. This means in particular that the classical part of the conductivity, which stems from the diagonal terms, remains unaffected by disorder. This behavior is not surprising, since it just reflects our underlying assumption, namely that the transport MFP l_T is much longer than the trajectories involved. In order to obtain the disorder damping of the conductivity one has to consider trajectories of the order of or longer than l_T. This implies that one allows for a deformation of the clean trajectories due to small-angle scattering in the impurity potential. Mirlin et al. [284], following such lines, indeed find within a related semiclassical approach a damping of the two-point Green function on the scale of l_T. To this end they have to start from a product of time-dependent propagators and perform Laplace transformations (to obtain energy-dependent Green functions) simultaneously, going beyond the stationary-phase approximation. Their approach works for the bulk but, to our knowledge, cannot yet be generalized to (periodic) orbits in confined systems.

In the opposite case, where two trajectories t, t' in (5.20) are completely uncorrelated, the disorder average in (5.20) factorizes: $\langle G_t(E)\, G_{t'}^*(E)\rangle = \langle G_t(E)\rangle \cdot \langle G_{t'}^*(E)\rangle$, leading to single-particle damping behavior. This holds typically for (long) trajectories in classical chaotic systems. Therefore nondiagonal contributions are exponentially suppressed in the presence of disorder. A proper semiclassical treatment of these nondiagonal terms in clean systems is an outstanding problem, as discussed in Sects. 2.2 and 3.2.1.

Orbits with a spatial separation larger than ξ are exponentially damped in integrable systems, too. However, different orbits from the same family of an integrable system which stay within a distance $\leq \xi$ are affected by smooth disorder in a *correlated* way. In this case the behavior of $\langle G_t(E)\, G_{t'}^*(E)\rangle$ is more complicated and depends on the confinement geometry. The damping then typically shows an algebraic behavior [72, 283] which will be illustrated in Sect. 5.4.2 for the case of the disorder-averaged magnetic susceptibility.

5.3 High Landau Levels in a Smooth Disorder Potential

As the first example of impurity effects on a quantity based on a single-particle Green function we briefly discuss the broadening of Landau levels in a smooth random potential. This issue has received ongoing interest in the past, owing in particular to its relation to the quantum Hall effect. The earlier literature on this problem was reviewed in detail by Ando et al. [49]. A more recent summary of the related theoretical approaches, including exact solutions for short-range disorder models, the use of the instanton method and calculations in the framework of the self-consistent Born approximation, can be found in [288].

Here we address the effect of a *smooth* random potential ($\xi \gg \lambda_F$) on the density of states and, hence, on the related de Haas–van Alphen oscillations in the magnetization.

To this end we start from the semiclassical expression for the diagonal part of the Green function for electrons in a homogeneous magnetic field H in the disorder-free case. It is a sum over contributions from cyclotron orbits and their repetitions j and reads, omitting the Weyl part (according to (4.90) and (4.91)),

$$G(r, r'=r; E) = \sum_{j=1}^{\infty} (-1)^j \frac{m}{\hbar^2} \exp\left(ij\frac{S_{cyc}}{\hbar} - i\frac{\pi}{2}\right) . \tag{5.21}$$

Here $S_{cyc} = 2\pi E/\hbar\omega_{cyc} = \pi k r_{cyc}$ is the action of a cyclotron orbit of radius r_{cyc}, and frequency ω_{cyc} in the clean system. In the presence of a random impurity potential each cyclotron orbit will be exponentially damped according to (5.9). Upon taking the trace of the Green function and reincluding the Weyl part we then find for the impurity-averaged density of states

$$d(E, H) = g_s \frac{mA}{2\pi\hbar^2}$$

$$\times \left[1 + 2\sum_{j=1}^{\infty} (-1)^j \cos\left(\frac{2\pi jE}{\hbar\omega_{cyc}}\right) \exp\left(-j^2 \frac{\langle \delta S_{cyc}^2 \rangle}{2\hbar^2}\right)\right] \tag{5.22}$$

for a sample of area A. In this particular case of cyclotron motion the integral (5.10) for the semiclassical damping term can be exactly evaluated. One finds, for Gaussian disorder (see (5.6)),

$$\langle \delta S_{cyc}^2 \rangle = \frac{2\pi n_i u^2}{v^2} f(\kappa) \tag{5.23}$$

with

$$f(\kappa) \equiv \kappa \, e^{-\kappa} I_0(\kappa) . \tag{5.24}$$

I_0 is a modified Bessel function. In the above equation κ depends on the ratio between the cyclotron radius and the correlation length and can alternatively be expressed by means of the magnetic length

$$L_H = \sqrt{\phi_0/2\pi H} \ : \tag{5.25}$$

$$\kappa = \frac{1}{2}\left(\frac{r_{cyc}}{\xi}\right)^2 = \frac{1}{2}(L_H k)^2 \left(\frac{L_H}{\xi}\right)^2 . \tag{5.26}$$

Using the Poisson summation formula one obtains, after a few manipulations [289],

$$d(E,H) = \frac{g_s A}{2\pi L_H^2} \sum_{k=0}^{\infty} \frac{(-1)^k}{\sqrt{2\pi}\Gamma(E)} \exp\left[-\frac{(E-E_k)^2}{2\Gamma^2(E)}\right] \qquad (5.27)$$

with $E_k = (k+1/2)\hbar\omega_{cyc}$. The density of states is given as a sum of Landau levels with Gaussian (energy-dependent) widths of the form

$$\Gamma^2(E) = \frac{n_i u^2}{2\pi r_{cyc}^2}\ f(\kappa) = \frac{1}{\sqrt{\pi}}\frac{\xi}{l}(\hbar\omega_{cyc})^2 f(\kappa). \qquad (5.28)$$

Here, l is the elastic MFP for smooth disorder (see (A.45) and (A.46)).

In the limit $r_{cyc} \gg \xi$ we obtain from the asymptotic form of I_0

$$\Gamma^2(E) = \frac{(\hbar\omega_{cyc})^2}{2\pi}\frac{r_{cyc}}{l}. \qquad (5.29)$$

In this case the width is proportional to $E^{-1/4}$ since l scales as E and r_{cyc} is proportional to $E^{1/2}$. This decrease of the width with energy results from the effective averaging out of the random potential fluctuating along the cyclotron orbit. The field dependence of the width is $\Gamma \sim \sqrt{H}$.

In the opposite limit $r_{cyc} \ll \xi$ of ultrasmooth disorder the width is

$$\Gamma^2 = \frac{(\hbar\omega_{cyc})^2}{\sqrt{4\pi}}\frac{r_{cyc}^2}{l\xi}. \qquad (5.30)$$

The change of the disorder potential along the cyclotron orbit is negligibly small and the width Γ of the Landau levels is independent of energy and the level index k. It is also independent of the H field. In this case the (inhomogeneous) broadening arises from the average of cyclotron orbits located at regions which differ in the height of the random potential.

The disorder broadening of Landau levels is often addressed using quantum diagrammatic approaches. In the case of δ-correlated disorder, calculations in the frame of the self-consistent Born approximation (SCBA) usually rely on diagrams without self-intersections [49]. These approaches give the known semielliptic shape of the Landau levels but, for example, do not reproduce the Gaussian tails of Wegner's exact solution [290] for the lowest Landau level. The effect of a smooth random potential on high Landau levels was recently studied by Raikh and Shahbazyan [288] on the basis of quantum diagrammatic perturbation theory. These authors point out that for $\xi \geq L_H$ diagrams with self-intersections have to be considered. For $\xi \gg L_H$ these authors could perform the summation of all the diagrams and obtained exactly (5.27). The present semiclassical treatment of the disorder broadening allows a physically transparent and short independent derivation of this result. The Gaussian shape is simply related to the fact that the disorder dephasing enters with the square of the number of repetitions along a prime cyclotron orbit.

The field derivative of the expression (5.22) for the single-particle density of states gives directly the susceptibility of a bulk 2DEG in a uniform magnetic field in the presence of a smooth disorder potential:

$$\frac{\chi^{\mathrm{GC}}}{\chi_{\mathrm{L}}} = -1 - 24 \left(\frac{\mu}{\hbar\omega_{\mathrm{cyc}}} \right)^2 \tag{5.31}$$

$$\times \sum_{j=1}^{\infty} (-1)^j \cos\left(\frac{2\pi j\mu}{\hbar\omega_{\mathrm{cyc}}} \right) R\left(\frac{j2\pi r_{\mathrm{cyc}}}{L_T} \right) \exp\left[-j^2 \frac{2\pi^2 \Gamma(\mu)^2}{(\hbar\omega_{\mathrm{cyc}})^2} \right].$$

In the above equality we have additionally included the semiclassical temperature damping function R (2.49). The exponent Γ of the disorder damping is given in (5.28).

The magnetization of a two-dimensional bulk heterostructure was measured by Eisenstein et al. [248].[2] This experiment shows a Gaussian broadening of the Landau levels with a $\sqrt{H}$ dependence of the widths in complete accordance with the above semiclassical expressions and the diagrammatic results of [288].

We note that the present semiclassical approach to the de Haas–van Alphen oscillations has been further extended in [284] to treat the corresponding transport quantity, namely the Shubnikov–de Haas oscillations of the diagonal magnetoconductivity.

5.4 Magnetic Susceptibility of Ballistic Quantum Dots

In the following we shall use the semiclassical approach to disorder averaging in finite systems developed in Sect. 5.2 to study impurity effects on the susceptibility of quantum dots in the ballistic regime [72, 283]. When discussing ensemble averaging for quantum dots one has to distinguish two situations:

(i) A disorder average of an ensemble of structures for which the parameters of the corresponding clean system, such as their geometry, size, and chemical potential, remain fixed while considering different impurity realizations. This kind of average will be henceforth called a *fixed-size impurity average* and will be studied below.
(ii) A *combined* energy (or size) and disorder average, which is experimentally more relevant and will be treated in Sects. 5.4.2 and 5.5.3.

5.4.1 Fixed-Size Impurity Average

As shown in Sect. 5.2, oscillatory contributions to the single-particle Green function from paths reflected at the boundaries of clean confined systems are exponentially damped but, depending on the elastic MFP, are not completely suppressed upon disorder averaging. As an observable quantity depending on these contributions we first consider the impurity-averaged susceptibility $\langle \chi \rangle$

---

[2] For a recent magnetization measurement in the fractional quantum Hall regime see [249].

of an ensemble of billiards of the same size or of same clean-system Fermi energy.

Disorder effects are expected to be most significant in systems with a regular clean counterpart where the weak random potential perturbs the integrable dynamics. We shall therefore focus on regular billiards at zero field or small magnetic fields, where the integrability is approximately maintained. In this case the grand canonical contribution $\chi^{(1)}$ to the susceptibility of a noninteracting clean structure is given by (4.49) and (4.53). In the presence of disorder this general result formally persists up to the replacement of the field-dependent part $\mathcal{C}_{\mathbf{M}}(H)$ (see (4.49)) by

$$\langle \mathcal{C}_{\mathbf{M}}(H) \rangle = \frac{1}{2\pi} \int_0^{2\pi} d\Theta_1 \, \cos\left[2\pi \frac{H \mathcal{A}_{\mathbf{M}}(\Theta_1)}{\phi_0} \right] \exp\left[-\frac{\langle \delta S_{\mathbf{M}}^2(\Theta_1) \rangle}{2\hbar^2} \right] . \quad (5.32)$$

The damping exponent $\langle \delta S_{\mathbf{M}}^2(\Theta_1) \rangle$ is given by (5.10). The integrals therein are performed along the orbits of the family $\mathbf{M}$, which are parametrized by Θ_1.

In the finite-range case, $a > \xi > \lambda_{\mathrm{F}}$, the disorder damping depends solely on the orbit length. If, as in the case of billiards, all orbits of a family $\mathbf{M}$ are of the same length, the family exhibits a unique disorder damping, giving a susceptibility contribution

$$\langle \chi_{\mathbf{M}}^{(1)} \rangle = \chi_{\mathbf{M}}^{(1)} \exp\left(-\frac{\langle \delta S_{\mathbf{M}}^2 \rangle}{2\hbar^2} \right) = \chi_{\mathbf{M}}^{(1)} \exp\left(-\frac{L_{\mathbf{M}}}{2l} \right) . \quad (5.33)$$

$\chi_{\mathbf{M}}^{(1)}$ is the contribution of the family $\mathbf{M}$ to the susceptibility (4.53) without disorder. In the long-range case $a < \xi$ the damping $\langle \delta S_{\mathbf{M}}(\Theta_1))^2 \rangle$ generally depends on Θ_1.

As a specific example we shall refer again to the case of square billiards. As discussed in Sect. 4.5, the dominant susceptibility contribution at finite temperature stems from the family $\mathbf{M} = (1,1)$ shown in the inset of Fig. 5.1. x_0 labels the different orbits of the family. In the presence of disorder the susceptibility reads, in analogy with (4.70) [72],

$$\frac{\langle \chi \rangle}{\chi^0} \simeq \frac{\langle \chi_{11}^{(1)} \rangle}{\chi^0} \quad (5.34)$$

$$\simeq \int_0^a \frac{dx_0}{a} \mathcal{A}^2(x_0) \cos[\varphi \mathcal{A}(x_0)] \left\langle \sin\left(k_{\mathrm{F}} L_{11} + \frac{\pi}{4} + \frac{\delta S(x_0)}{\hbar} \right) \right\rangle$$

with χ^0 given by (4.71).

In the finite-range regime, $\delta S(x_0)$ is the same for all orbits $(1,1)$ and the susceptibility is exponentially damped according to (5.33) with an exponent $L_{11}/2l$. Moreover, in the specific case of a square billiard $\delta S(x_0)$ turns out to be also independent of x_0 in the opposite long-range regime. Again, the contribution from family $(1,1)$ exhibits a unique exponential damping, given by (5.19). In this case I_{11} entering into the damping exponent is the same for all orbits $(1,1)$, i.e. $I_{11} = a^2/12$.

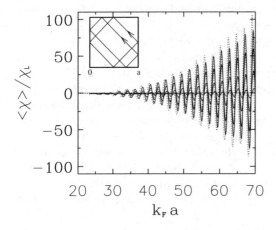

Fig. 5.1. Magnetic susceptibility $\langle \chi \rangle$ (normalized with respect to the Landau susceptibility χ_L) of a square billiard as a function of $k_F a$ for the clean case (*dotted*) and for the ensemble average of billiards of fixed size with increasing Gaussian disorder ($\xi/a = 0.1$) according to an elastic mean free path $l/a = 4, 2, 1, 0.5$ (*solid lines* in order of decreasing amplitude). $\langle \chi \rangle$ is calculated for $H = 0$ and $k_B T = 6\Delta$. The *inset* shows two representative orbits of family (1,1). (From [72], by permission)

As a test of the analytical approximations presented above, we shall compare the semiclassical results with numerical quantum calculations described in Appendix A.2. These are based on a diagonalization of the Hamiltonian for noninteracting particles in a square billiard subject to a uniform perpendicular H field and a random disorder potential of the form of (5.5). For a given selected correlation length ξ, elastic MFP l_{qm}, and fixed Fermi momentum k_F, the product $n_i u^2$ is determined from (A.44) and (A.45) in Appendix A.3 and used as am entry in the quantum calculation.

For an *individual* square billiard and $\xi < a$ each impurity configuration i has a self-averaging effect. This arises from the differences in the impurity potential $V_i^{dis}(r)$ across the structure. In an average over an *ensemble* of square billiards, an additional damping appears owing to differences in the mean impurity potential $\overline{V_i^{dis}} = (1/a^2) \int dr\, V_i^{dis}(r)$ between different squares.

This damping effect is characterized by the variance (where $\eta = a/2\xi$):

$$\langle \overline{V^{dis}}^2 \rangle = \frac{u^2 n_i}{a^2 \eta^2} \left[\eta\, \mathrm{erf}(\eta) + \frac{1}{\sqrt{\pi}} \left(e^{-\eta^2} - 1 \right) \right]^2$$

$$\longrightarrow \begin{cases} u^2 n_i / 4\pi\xi^2 & \text{for} \quad \xi/a \longrightarrow \infty \\ u^2 n_i / a^2 & \text{for} \quad \xi \longrightarrow 0 \end{cases} . \tag{5.35}$$

In the limit of $\xi \gg a$ variations in the mean potential $\overline{V^{dis}}$, according to the above equation, dominate the damping: in this case the self-averaging effect is negligible because the impurity potential is essentially flat across each single square and the clean susceptibility of an individual structure remains practically unaffected by disorder.

In the opposite limit of white-noise disorder, the quantum calculations show [72] that sample-to-sample fluctuations in the mean $\overline{V}^{\mathrm{dis}}$ are of minor relevance. Self-averaging is the predominant process for an integrable system: in the semiclassical picture different trajectories within the same family of closed orbits are perturbed in an uncorrelated manner by white-noise disorder. Hence, for $\xi \ll a$ there are no significant differences between the susceptibility of an individual disordered billiard of integrable geometry and that of the corresponding ensemble. On the other hand, in a chaotic geometry this self-averaging effect does not exist since orbits are isolated (for ξ not too small, see end of Sect. 5.4.2). Thus in the case of chaotic billiards, distinct differences between a single weakly disordered sample and an ensemble with ballistic disorder are expected.

The results of numerical quantum simulations for the average susceptibility $\langle \chi \rangle$ of an ensemble of squares with fixed size but different disorder realizations are shown in Fig. 5.1. The dotted line represents the disorder-free case, showing the characteristic oscillations of χ in $k_{\mathrm{F}}a$ which are dominated by contributions from the paths of family (1,1) (see Sect. 4.5.2). The other curves in the figure show the damping of the clean susceptibility with decreasing elastic MFP $l/a = 4, 2, 1, 0.5$ for a fixed value of the correlation length $\xi/a = 0.1$, typical of experimental realizations. We note that even for $l \sim a$ the signature of the orbits (1,1) from the clean geometry persists.

A quantitative comparison of the disorder damping obtained from numerical and analytical results is depicted in Fig. 5.2: the logarithm of $\langle \chi \rangle$ normalized to the corresponding zero-disorder susceptibility χ_c is plotted as a function of the inverse MFP for different correlation lengths ξ. The straight lines represent the semiclassically predicted exponential damping according to (5.33) for the short-range regime $\xi = 0$ (full line) and according to (5.19) for the long-range regime $\xi > a$ (dotted lines for $\xi/a = 4, 2, 1$ from the top). The semiclassical predictions accurately agree with the corresponding quantum results (symbols) for $\xi/a = 4, 2, 1$ and show small deviations for $\xi = 0$. They fail for intermediate correlation values $\xi/a = 0.5, 0.2$ (squares and diamonds) which are outside the range of validity of the approximations used. We note that the crossover from self-averaging-dominated damping in the limit $\xi \to 0$ to a damping due to fluctuations in the mean $\overline{V}^{\mathrm{dis}}$ for $\xi/a \to \infty$ turns out to be nonmonotonic.

5.4.2 Combined Impurity and Size Average

For the samples used in current experiments, disorder averages cannot be performed independently from size averages. For instance, in an array of lithographically defined semiconductor microstructures the individual systems typically vary in size and shape, at least on small scales. The basic expression (5.34) for the susceptibility shows that changes in the size a lead

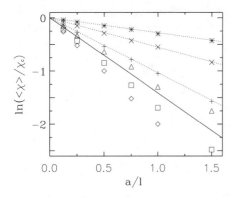

Fig. 5.2. Logarithm of the ratio $\langle\chi\rangle/\chi_{\rm c}$ as a function of the inverse elastic MFP a/l. The *symbols* indicate the numerical quantum results (from the *top* for $\xi/a = 4, 2, 1, 0, 0.5$, and 0.2). The *dotted lines* show the semiclassical analytical results for $\xi/a = 4, 2, 1$ (from the *top*) according to (5.19). The *full line* is the semiclassical result for $\xi = 0$ (see (5.12)). The quantum results for $\xi = 0.5$ (*squares*) and 0.2 (*diamonds*) are beyond the regime of validity of the analytical limits $\xi/a \gg 1$ and $\xi/a \ll 1$. (From [72], by permission)

to rapid variations in the phases $\sim k_{\rm F}a$ and are equivalent to an energy or $k_{\rm F}$ average. We therefore address in the following the energy- and impurity-averaged susceptibility $\langle\overline{\chi}\rangle$ describing the magnetic response of an ensemble of microstructures with different impurity realizations and variations in size. This describes, for instance, the experimental setup reported in [57].

As discussed in the context of clean systems, an energy average leads to the vanishing of the oscillations from $\chi^{(1)}$ shown, for example, in Fig. 5.1. Instead, canonical corrections $\chi^{(2)}$, which are related to the number variance according to (4.24b), have to be considered. This variance is in turn related to $K(\varepsilon_1, \varepsilon_2; H)$ (see (5.1)) by integration of the level energies ε_1, ε_2 over the energy interval. According to the decomposition $K = \overline{K^{\rm d}} + K^{\rm s}$, the total disorder- and size-averaged susceptibility is composed of

$$\overline{\langle\chi(H)\rangle} = \overline{\langle\chi^{\rm d}(H)\rangle} + \overline{\langle\chi^{\rm s}(H)\rangle}. \tag{5.36}$$

In the following we study the contribution $\overline{\langle\chi^{\rm s}(H)\rangle}$ and focus again on integrable structures. In order to obtain the semiclassical result for $\overline{\langle\chi^{\rm s}(H)\rangle}$, disorder averaging is incorporated into the formulas (4.55) for the disorder-free case in an analogous way to that for $\langle\chi\rangle$ in Sect. 5.4.1. One has to include in the integral (4.56) for $\mathcal{C}_{\rm M}''$ a Θ_1-dependent disorder-induced phase $\exp(\mathrm{i}\delta S(\Theta_1)/\hbar)$. However, since we have to take the square of $\mathcal{C}_{\rm M}''$ before the impurity average, cross correlations between different paths Θ_1 and Θ_1' on the same torus **M** or between different tori have to be considered.

We discuss this effect, typical of integrable systems, again for the case of the square billiard. For the sake of clarity we furthermore assume temperatures such that only the contribution of the shortest closed orbits $(1,1)$

has to be considered. Instead of (4.74) for the clean case, the contribution of these orbits to the energy- *and* disorder-averaged susceptibility is now of the form [72]

$$\frac{\langle \overline{\chi^s} \rangle}{\overline{\chi^0}} = \frac{1}{2} \int_0^a \int_0^a \frac{dx_0 \, dx_0'}{a} \left[\mathcal{A}_-^2 \cos(\varphi \mathcal{A}_-) + \mathcal{A}_+^2 \cos(\varphi \mathcal{A}_+) \right] f(x_0, x_0') \quad (5.37)$$

with

$$\frac{\overline{\chi^0}}{\chi_L} = \frac{3}{(\sqrt{2}\pi)^3} \, (k_F a) \, R^2 (L_{11}/L_T) \,. \quad (5.38)$$

$\mathcal{A}_\pm = \mathcal{A}(x_0) \pm \mathcal{A}(x_0')$ and $\mathcal{A}(x_0) = 4\pi x_0 (a - x_0)/a^2$. The correlation function

$$f(x_0, x_0') = \left\langle \exp\left\{ \frac{i}{\hbar} \left[\delta S(x_0) - \delta S(x_0') \right] \right\} \right\rangle \quad (5.39)$$

$$= \exp\left\{ -\frac{1}{2\hbar^2} \left[\langle \delta S^2(x_0) \rangle + \langle \delta S^2(x_0') \rangle - 2\langle \delta S(x_0) \delta S(x_0') \rangle \right] \right\}$$

accounts for the effect of disorder on pairs of trajectories x_0 and x_0'. The disorder correlation length ξ, entering implicitly into f, leads to characteristic features of the smooth-disorder damping as will be discussed below.

For *short-range disorder*, $\xi \ll \lambda_F$, we reach the border of applicability of the semiclassical approach. However, the semiclassical approach shows that for $\xi \to 0$ orbits with $x_0 \neq x_0'$ remain disorder-uncorrelated. Hence, the corresponding pair contributions are exponentially suppressed. This leads to an overall damping of the average susceptibility, which reads (at finite temperature, on the basis of the family (1,1))

$$\lim_{\xi \to 0} \langle \overline{\chi^s} \rangle = \overline{\chi} \, e^{-L_{11}/l_\delta} \,. \quad (5.40)$$

The exponent for $\langle \overline{\chi^s} \rangle$ differs by a factor $1/2$ from that for $\langle \overline{\chi} \rangle$ given by (5.33). The $k_F a$ dependence of the ensemble-averaged susceptibility $\langle \overline{\chi^s} \rangle$, together with $\langle \overline{\chi^d} \rangle$, will be compared with numerical quantum results in Sect. 5.5.3.

In the *finite-range case*, $\lambda_F < \xi \ll a$, the phase shifts $\delta S(x_0)$ and $\delta S(x_0')$ entering into $f(x_0, x_0')$ are accumulated along the orbits in a correlated way, if the spatial distance between the two orbits x_0 and x_0' is smaller than ξ. In this regime the product term $2\langle \delta S(x_0) \delta S(x_0') \rangle$ in the exponent of $f(x_0, x_0')$ is given by integrals as in (5.10) but with $\boldsymbol{q}$ and $\boldsymbol{q}'$ running along the different paths. Two trajectories x_0 and x_0' of a family of orbits in a billiard can be regarded as parallel straight lines with a distance between them of $y = |x_0 - x_0'|/\sqrt{2}$ if one neglects the additional correlations occurring near the reflections at the boundaries. Then the integral of type (5.10) can be evaluated analytically and we have [72]

$$f(x_0, x_0') = \exp\left\{ -\frac{L_{11}}{l} \left[1 - \exp\left(-\frac{(x_0 - x_0')^2}{8\xi^2} \right) \right] \right\} \,. \quad (5.41)$$

Contributions from disorder-correlated trajectories separated by $|x_0 - x_0'| < \xi$ are only weakly damped. In contrast, for orbits separated by $|x_0 - x_0'| \gg \xi$

individual random disorder leads to an uncorrelated detuning of the phases. Thus they are exponentially suppressed: $f(x_0, x_0') \simeq \exp(-L_{11}/l)$.

In general, disorder averages in the finite-range regime lead, by means of the correlation function f, to a *nonexponential* damping of the susceptibilities for systems with families of periodic orbits.

We study this behavior more explicitly for the case of square billiards where the corresponding integral (5.37) can be evaluated analytically at $H = 0$ in the limits of $L_{11} \ll l$ and $L_{11} \gg l$. The size- and disorder-averaged susceptibility at $H = 0$, for example, reads in the limit $L_{11} \gg l$ [72],

$$\frac{\overline{\langle \chi^s \rangle}}{\overline{\chi}} \simeq 2\sqrt{2\pi} \left(\frac{\xi}{a} \right) \left(\frac{l}{L_{11}} \right)^{1/2} . \tag{5.42}$$

This limit is particularly interesting since (5.42) expresses the fact that disorder correlation effects lead to a replacement of the exponential damping by a power law.

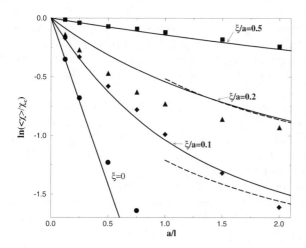

Fig. 5.3. Logarithm of the ratio between disorder-averaged and clean results for ensemble-averaged $\langle \chi^s \rangle$ susceptibilities as a function of increasing inverse elastic MFP a/l for different values of ξ/a. The *symbols* denote the numerical quantum results, the *solid lines* (for $\xi > 0$) the semiclassical integrals (5.37), and the *dashed lines* asymptotic expansions (5.42) of the integrals for large a/l. (From [283]; ©1996 by the American Physical Society)

The numerical quantum mechanical and analytical semiclassical results for the disorder averaged susceptibility of ensembles of square billiards are summarized in Fig. 5.3 (at $H = 0$ and $k_B T = 2\Delta$). It depicts, in semilogarithmic representation, $\overline{\langle \chi^s \rangle}$ as a function of the inverse elastic MFP for different disorder correlation lengths $\xi/a = 0.5, 0.2, 0.1, 0$ (from the top). The symbols denote the quantum results from the simulations described in the previous section. The full curves are semiclassical results from numerical integration of (5.37). In the short-range limit $\xi = 0$ they reduce to (5.40) and give an exponential decrease with exponent L_{11}/l. This damping is also found from the quantum simulations (circles).

The analytical expression (5.42), valid for $L_{11} > l$, is represented as dotted curves in the limit $a/l \geq 1$. This regime can still be considered as (deep) ballistic since for finite ξ the transport MFP $l_T > l$ (see (A.47)). Therefore, the semiclassical perturbative approach based on straight-line trajectories remains justified. A similar overall behavior is found in [72] for the typical susceptibility $\chi^{(t)} = \langle \overline{\chi^2} \rangle^{1/2}$.

5.4.3 Concluding Remarks

As pointed out, the interplay between confinement and a smooth random potential leads to peculiar disorder-damping effects in quantum dots. The need to consider different ensemble averages is inherent in ballistic nanostructures which are sufficiently small, and therefore usually nonself-averaging, contrary to the bulk. The different types of impurity averages studied in the previous sections are summarized in Table 5.1 for the magnetic susceptibility of integrable systems. The table distinguishes between the three regimes defined by the correlation length of the impurity potential. The results can be summarized as follows.

The fixed-size averaged susceptibility of an integrable structure, obtainable from the average of one-particle Green functions, is always exponentially damped by disorder. In the short-range regime ($\xi < \lambda_F$) the damping is governed by the length L of the most relevant trajectories and the elastic MFP l_δ. This result persists in the finite-range regime ($\lambda_F < \xi \ll a$), but with an elastic MFP evaluated semiclassically. In the long-range regime ($\xi > a$) the fixed-size averaged susceptibility depends exponentially on the product $(L/l)(L/\xi)$ and a correction containing the geometry of the periodic trajectories.

For comparison with actual experiments one has to consider that different impurity realizations are accompanied by a change in the Fermi energy and the size of the structure. Therefore impurity- and size-averaged susceptibilities, which are expressed by two-particle Green functions, are most relevant. For the short-range case the only difference between one- and two-point Green function quantities is the factor $1/2$ of the exponential damping of the former. However, in the finite-range regime there appear important differences: pairs of closed trajectories that remain at distances smaller than the correlation length ξ give rise to a weak damping with a power-law dependence on l/L and ξ/a for integrable microstructures. The disorder damping is affected decisively by finite-size effects since it depends not only on bulk-like characteristics of the disorder like the elastic MFP, but also on the ratio between the size of the structure and the correlation length of the potential.

Table 5.1. Summary of the different average susceptibilities considered in the short-range ($\xi < \lambda_F < a$), finite-range ($\lambda_F < \xi < a$), and long-range ($\lambda_F < a < \xi$) regimes at $H = 0$. The fixed-size impurity-averaged susceptibility $\langle\chi\rangle$ is given by the one-particle Green function; the average susceptibility $\overline{\langle\chi^s\rangle}$ is related to two-particle Green functions involving both impurity and size averages. The different average susceptibilities are normalized with respect to the corresponding quantities of the disorder-free structures. The numerical factors are (for the square) $c_2 = 2\sqrt{2\pi}, d_1 = 1/4\sqrt{\pi}$, and $d_2 \simeq 6.5 \times 10^{-5}$. (From [283])

	Short-range	Finite-range	Long-range
$\langle\chi\rangle/\chi$	$\exp\left(-L_{11}/2l_\delta\right)$	$\exp\left(-L_{11}/2l\right)$	$\exp\left\{-d_1(L_{11}^2/l\xi)[1-I_t/(2\xi^2)]\right\}$
$\overline{\langle\chi^s\rangle}/\bar{\chi}$	$\exp\left(-L_{11}/l_\delta\right)$	$c_2(\xi/a)(l/L_{11})^{1/2}$	$1 - d_2a/l(a/\xi)^9$

The disorder damping discussed in the previous subsections is characteristic of integrable geometries. The main damping effect in the finite-range regime is related to a self-averaging mechanism: the random disorder phases accumulated along the orbits of a family destroy the constructive interference from in-phase contributions from all orbits of the family in a clean integrable system. Hence, the large susceptibility of an ensemble of clean square billiards is reduced accordingly. The disorder correlation length determines the spatial regions over which trajectories within a family still accumulate a correlated phase.

In clean chaotic geometries, periodic trajectories are usually isolated, resulting in smaller oscillations of the density of states and a much weaker magnetic response than in integrable systems. Introduction of disorder in chaotic geometries is therefore less dramatic than in integrable systems, since it merely changes the action of the relevant periodic trajectories instead of producing dephasing within a family. Self-averaging effects do not exist.

At the end of the following section we shall compare disorder effects in integrable and chaotic geometries and make contact with experiment after a discussion of the disorder-induced spectral correlation K^d and related magnetic response $\overline{\langle\chi^d\rangle}$.

5.5 From Ballistic to Diffusive Dynamics

For the weak disorder in the ballistic regime that we have considered so far in this chapter, the lowest-order approximation consists of the perturbative modification of the classical actions by the impurity potential, keeping the trajectories of the clean geometry. This modification gives rise to the size-induced correlations K^s to the spectral two-point correlator $K(\varepsilon_1, \varepsilon_2; H)$ (5.1). With increasing disorder, however, both K^s and disorder-induced correlations K^d, contributions from scattering between impurities, have to be considered. This is the aim of this section.

5.5.1 Spectral Correlations in the Diffusive Limit

We begin with the diffusive limit $l \ll a$ (of white-noise disorder), where periodic-orbit contributions to K^s are strongly suppressed. In this limit, the correlations K^d are dominant. The treatment of the disordered *bulk* by means of diffusive trajectories was comprehensively reviewed by Chakravarty and Schmid [7]. These authors showed that several quantum phenomena, such as weak localization, can be understood in terms of coherent quantum superposition of waves traveling along classical paths and being scattered at the impurities.

A semiclassical analysis of spectral correlations of *finite* diffusive systems with an interpretation of the diagrams in terms of classical paths was later put forward by Argaman, Imry, and Smilansky [76]. These authors expressed the spectral form factor $\tilde{K}(E,t)$ (2.31), the Fourier transform of the two-point correlation function, through classical return probabilities $P(t)$ and found $\tilde{K}(E,t) \sim |t| P(t)$ (see (4.46) and related discussion). The form factor, which is sketched in Fig. 5.4, reveals different mesoscopic behaviors in various time regimes. For timescales smaller than the ergodic or Thouless time,

$$t_{\mathrm{erg}} \sim \frac{a^2}{D} \tag{5.43}$$

(D is the diffusion constant), the return probability is governed by diffusive dynamics: $P(t) \sim (Dt)^{-N/2}$ in N dimensions and hence $\tilde{K}(E,t) \sim t^{1-N/2}$. For times larger than t_{erg} the dynamics is assumed to be ergodic and the return probablility is constant. Therefore $\tilde{K}(E,t) \sim t$. For times larger than the Heisenberg time $T_{\mathrm{H}} = 2\pi\hbar\bar{d}$ the form factor saturates in the quantum regime. A related semiclassical analysis in the same spirit has been reviewed by Dittrich for one-dimensional disordered systems [77].

5.5.2 Spectral Correlations in Disordered Nondiffusive Systems

The scenario depicted in Fig. 5.4 holds true for the diffusive regime where the mean free time τ is small compared to t_{erg} or $l \ll a$. In the following we review an approach [282] to finite disordered *nondiffusive* systems, where the usual diagrammatic techniques for treating impurity scattering are not directly applicable. The combination of a diagrammatic perturbation approach with semiclassical techniques allows us to calculate the disorder-induced part K^d (5.2a) of the energy correlation function. In this quantum-semiclassical hybrid approach, scattering at the impurity potentials, which are now assumed to be δ-like, is treated quantum mechanically in a perturbation series with respect to the disorder. Boundary effects are incorporated in a semiclassical representation of the Green functions, which enter into the impurity

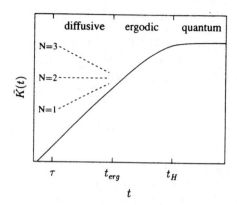

Fig. 5.4. Sketch of the spectral form factor. The *full line* represents the orthogonal random matrix result and the *dashed lines* the modifications for N-dimensional diffusive systems. The different time regimes (diffusive, ergodic, quantum) are classified by the following transition times: the mean free time τ, the ergodic time $t_{\text{erg}} = a^2/D$, and the Heisenberg time $T_{\text{H}} = 2\pi\hbar\bar{d}$. (From [76], by permission; ©1993 by the American Physical Society)

diagrams, in terms of classical paths.[3] This method, therefore, takes into account in a systematic way contributions from (closed) trajectories which involve both scattering at impurities and specular reflection at the confinement potential. The approach, presented here for billiard systems, applies also to systems with potentials whose effect is then incorporated into the semiclassical Green functions. Together with the size-induced correlations (5.2b), which stem from (damped) periodic-orbit contributions and were considered in Sects. 5.2–5.4, the procedure allows us to study the complete crossover from diffusive to clean confined systems.

We note that the spectral statistics of the crossover regime between ballistic and diffusive dynamics have been previously addressed by Altland and Gefen [278]. These authors considered the case of a square with disorder and devised a diagrammatic treatment to go beyond the common diffusion and Cooperon approximation. A similar method was applied to cylinder geometries [295]. Agam and Fishman [296] placed finite hyperspherical scatterers in a torus-like system and studied the spectral form factor in the crossover regime in terms of dynamical zeta functions[4] of the associated classical system. A nonperturbative approach for ballistic systems has been developed by Muzykantskii and Khmelnitskii, the ballistic σ model [47]. It has been more recently applied to the case of a clean disk with rough boundaries [297].

---

[3] This approach has features of the "method of trajectories" originally devised for superconducting films [294]. A similar method will be presented in Chap. 6 to compute interaction effects in ballistic dots.

[4] See Sect. 2.2 for the role of zeta functions in the level correlator.

Diagrammatic Framework

We first review the diagrammatic ansatz used by Altland and Gefen [278] for disorder-induced correlations and then perform the semiclassical evaluation. We consider noninteracting electrons in a weak, perpendicular magnetic field and a random impurity potential V^{dis}, with a correlator (5.3) describing white-noise disorder:

$$\left\langle V^{\text{dis}}(\boldsymbol{r})\, V^{\text{dis}}(\boldsymbol{r}')\right\rangle = \frac{\hbar}{2\pi N(0)\tau}\, \delta(\boldsymbol{r}-\boldsymbol{r}')\,. \tag{5.44}$$

Here, τ is the mean elastic scattering time, $l \equiv l_\delta = v_F \tau$, and $N(0) = \bar{d}/A$. In terms of the retarded and advanced single-particle Green functions $G^{+(-)}(\boldsymbol{r}_1, \boldsymbol{r}_2; \varepsilon; H)$, obeying the boundary conditions of the corresponding clean system, the correlator K^{d} may be written as

$$K^{\text{d}}(\varepsilon_1, \varepsilon_2; H) \approx \left(\frac{\Delta^2}{2\pi^2}\right) \text{Re}\,\left\langle\!\!\left\langle \text{Tr}\, G^+(\varepsilon_1; H)\, \text{Tr}\, G^-(\varepsilon_2; H)\right\rangle\!\!\right\rangle\,. \tag{5.45}$$

At this stage, the average is taken for a given system size over impurities only. The symbol $\left\langle\!\!\left\langle \ldots \right\rangle\!\!\right\rangle$ implies the inclusion of connected diagrams only [282].

The correlator K^{d} can be expressed diagrammatically as [278]

$$K^{\text{d}}(\epsilon_1, \epsilon_2; H) = \frac{\Delta^2}{2\pi^2} \frac{\partial}{\partial\epsilon_1} \frac{\partial}{\partial\epsilon_2} \text{Re} \sum_{n=1}^{\infty} \frac{1}{n}\left[\mathcal{S}_n^{(\text{C})}(\omega; H) + \mathcal{S}_n^{(\text{D})}(\omega)\right] \tag{5.46}$$

with $\omega = \varepsilon_1 - \varepsilon_2$. We are particularly interested in the field-sensitive part given by the Cooperon-type diagrams $\mathcal{S}_n^{(\text{C})}$. These are defined by

$$\mathcal{S}_n^{(\text{C})}(\omega; H) = \text{Tr}\,\left[\zeta^{(\text{C})}(\omega; H)\right]^n \tag{5.47}$$

$$= \left(\int \prod_{j=1}^{n} \mathrm{d}^d r_j\right) \prod_{m=1}^{n} \zeta^{(\text{C})}(\boldsymbol{r}_m, \boldsymbol{r}_{m+1}; \omega; H)$$

with $\boldsymbol{r}_{n+1} \equiv \boldsymbol{r}_1$ and

$$\zeta^{(\text{C})}(\boldsymbol{r}_1, \boldsymbol{r}_2; \omega; H) = \frac{\hbar}{2\pi N(0)\tau} \left\langle G^+(\boldsymbol{r}_1, \boldsymbol{r}_2; \varepsilon_1; H)\right\rangle \left\langle G^-(\boldsymbol{r}_1, \boldsymbol{r}_2; \varepsilon_2; H)\right\rangle\,. \tag{5.48}$$

Here $\langle G^\pm\rangle$ is the disorder-averaged Green function. The diffuson-type diagrams $\mathcal{S}_n^{(\text{D})}$ are defined and can be evaluated correspondingly.

An example, $\mathcal{S}_4^{(\text{C})}$, is shown schematically in Fig. 5.5a. The sum of the diagrams $\mathcal{S}_n^{(\text{C})}$ yields the dominant contribution to the field-dependent part of K^{d} in the nondiffusive regime.

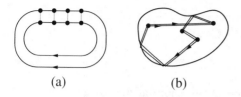

Fig. 5.5. Sketch of the diagrams $\mathcal{S}_n^{(C)}$: **(a)** shows a schematic form of $\mathcal{S}_4^{(C)}$, **(b)** shows a pair of typical real-space trajectories which contribute to $\mathcal{S}_4^{(C)}$. (From [282]; ©1999 by the American Physical Society)

Semiclassical Treatment

Owing to the lack of translational invariance in confined systems it is no longer convenient to evaluate diagrams such as that in Fig. 5.5 in momentum space. Instead we work in configuration space and compute the integrals (5.47) invoking a semiclassical approximation. Following [282] we use the semiclassical expression (5.17) for the impurity-averaged Green function. It is given as the sum over all classical paths t from $\boldsymbol{r}_1$ to $\boldsymbol{r}_2$,

$$\langle G^+(\boldsymbol{r}_1, \boldsymbol{r}_2; E)\rangle \simeq \sum_{t:\boldsymbol{r}_1 \to \boldsymbol{r}_2} \tilde{D}_t \exp\left(iS_t/\hbar - i\pi\eta_t/2 - L_t/2l\right), \qquad (5.49)$$

where $\tilde{D}_t = D_t/[2\pi(i\hbar)^3]^{1/2}$ (in two dimensions, with D_t given in (2.5)). This was derived in Sect. 5.2.2. Weak disorder leads to damping on the scale of l, leaving the trajectories unaffected, i.e. the sum is taken over the paths of the corresponding clean system.

Using the semiclassical expression (5.49) in (5.48), the two-particle operator $\zeta^{(C)}(\boldsymbol{r}_1, \boldsymbol{r}_2; \omega; H)$ is then given in terms of a double sum over pairs of classical paths. These pairs explicitly include the effect of boundary scattering. After energy or size averaging, the main contribution to the field-sensitive part of $\overline{K^d}$ arises from diagonal terms obtained by pairing paths with their time reverses (otherwise known as the Cooperon channel). For small H field we can expand the actions of the paths as

$$\frac{1}{\hbar}S_t(\epsilon_i; H) \simeq \frac{1}{\hbar}S_t(E_F; H{=}0) + (\epsilon_i - E_F)\tau_t + \frac{2\pi}{\phi_0}\int_{\boldsymbol{r}_1}^{\boldsymbol{r}_2} \boldsymbol{A}\cdot d\boldsymbol{r}, \qquad (5.50)$$

where τ_t is the period of the trajectory and $\boldsymbol{A}$ is the vector potential. Within the diagonal approximation $\zeta^{(C)}$ then reads

$$\zeta^{(C)}(\boldsymbol{r}_1, \boldsymbol{r}_2; \omega; H) = \sum_{t:\boldsymbol{r}_1 \to \boldsymbol{r}_2} \tilde{\zeta}_t^{(C)}(\boldsymbol{r}_1, \boldsymbol{r}_2; \omega; H), \qquad (5.51)$$

where

$$\tilde{\zeta}_t^{(C)}(\boldsymbol{r}_1, \boldsymbol{r}_2; \omega; H) \simeq \frac{v_F |\tilde{D}_t|^2}{2\pi N(0)l}$$

$$\times \exp\left(-\frac{L_t}{\ell_\phi} - \frac{L_t}{l} + i\omega\tau_t + i\frac{4\pi}{\phi_0}\int_{\boldsymbol{r}_1}^{\boldsymbol{r}_2} \boldsymbol{A}\cdot d\boldsymbol{r}\right). \qquad (5.52)$$

We have also introduced a level broadening via $\omega \to \omega + i\gamma$. The level broadening is implicit in the inelastic length ℓ_ϕ:

$$\frac{\ell_\phi}{a} = \frac{k_F a}{2\pi} \frac{\Delta}{\gamma} . \tag{5.53}$$

This may account either for damping of propagation due to inelastic scattering or temperature smearing. Equation (5.52) depends, apart from l, only on the system without disorder and holds for both integrable and chaotic geometries in the whole range from clean to diffusive.

The propagator $\zeta^{(C)}$ (5.51) is made up of a summation over all diagonal pairs of paths, including boundary scattering, between any two given impurities situated at r_1 and r_2. On taking the trace over n propagators $\zeta^{(C)}$, one sees, for example, that the field-sensitive part of S_n (5.47) consists of a summation over flux-enclosing pairs of closed paths (in position space) involving n impurities and an arbitrary number of boundary scattering events. An example of a pair of paths contributing to $\mathcal{S}_4^{(C)}$ is shown in Fig. 5.5b.

Representative Examples

The diffuson and Cooperon contributions to the disorder-induced correlator can be analytically computed only for certain confinement geometries. Here we present results for the disordered square quantum dot, representing integrable geometries, and for generic chaotic geometries. For the square we consider specular reflection at the boundaries and employ the extended zone scheme discussed in Sect. 4.5.1 (Fig. 4.11) to write $\zeta^{(C)}(r_1, r_2; \omega; H)$ (5.51) as a sum of propagators along *straight-line paths* $\tilde{\zeta}_t^{(C)}(r_1, r_2^t; \omega; H)$, of the form (5.52), where the r_2^t are images of the position r_2. The diagrams $\mathcal{S}_n^{(C)}$ (5.47) are then calculated by diagonalizing $\zeta^{(C)}$. At zero magnetic field this is analytically possible and one finds [278, 282]

$$\mathcal{S}_n^{(C,D)} = \sum_{m_x=0}^{\infty} \sum_{m_y=0}^{\infty} \left[\lambda \left(\frac{m_x \pi}{a}, \frac{m_y \pi}{a} \right) \right]^{-n} \tag{5.54}$$

with

$$\lambda(q_x, q_y) = \sqrt{(1 + \gamma\tau - i\omega\tau)^2 + l^2(q_x^2 + q_y^2)} . \tag{5.55}$$

Together with (5.46) this gives the spectral correlation function $\overline{K^d}$ of a disordered square for arbitrary l.

For systems with a generic chaotic, clean counterpart an analytical estimate of $\overline{K^d}$ can be achieved under certain statistical assumptions with respect to the classical trajectories involved. To this end we make use of the relation (2.7b) in order to transform the sums over classical densities $|\tilde{D}_t|^2$ in (5.51) into probabilities for propagating classically between impurities at r_1 and r_2 in time t. Let us assume that in the ballistic regime $l, \ell_\phi \gg a$ the conditional probability $P(\boldsymbol{r}_1, \boldsymbol{r}_2; t|\mathcal{A})$ to accumulate an "area" $\mathcal{A}$ during the propagation from $\boldsymbol{r}_1$ and $\boldsymbol{r}_2$ is independent of these points. Proceeding as in Sect. 4.3.2, where we considered clean billiards, we find, from (2.7b) and (5.51),

$$
\zeta^{(\mathrm{C})}(\boldsymbol{r}_1, \boldsymbol{r}_2; \omega; H) \simeq \frac{1}{\tau} \int_0^\infty dt \int_{-\infty}^\infty d\mathcal{A}\, P(\boldsymbol{r}_1, \boldsymbol{r}_2; t|\mathcal{A})
$$
$$
\times \cos\left(\frac{4\pi\mathcal{A}H}{\phi_0}\right) \exp\left(-\gamma t - \frac{t}{\tau} + i\omega t\right). \tag{5.56}
$$

We further assume that after several bounces off the boundary the area distribution $P(\boldsymbol{r}_1, \boldsymbol{r}_2; t|\mathcal{A})$ becomes Gaussian with a variance σ independent of l. Substituting the above expression for $\zeta^{(\mathrm{C})}$ into (5.47) and performing the r, t, and $\mathcal{A}$ integrals, we obtain the following closed approximate form for $\mathcal{S}_n^{(\mathrm{C})}$ [282]:[5]

$$
\mathcal{S}_n^{(\mathrm{C})}(\omega; H) \simeq \left(\frac{8\pi^2 H^2 l\sigma}{\phi_0^2} + 1 + \gamma\tau - i\omega\tau\right)^{-n}. \tag{5.57}
$$

The same result holds for the $\mathcal{S}_n^{(\mathrm{D})}$ with $H = 0$. Together, these yield the disorder-induced correlations for a chaotic geometry used in (5.46).

5.5.3 Orbital Magnetism

In the following we apply the above formalism to compute the magnetic response of ensembles of disordered billiards with regular and with chaotic geometry.[6] Employing (5.46) and the relation between $\overline{\langle\chi\rangle}$ and the density correlator (see (5.36) and related text in Sect. 5.4.2), the disorder-induced contribution to the average magnetic susceptibility is directly given by (with $\varphi = Ha^2/\phi_0$)

$$
\frac{\overline{\langle\chi^{\mathrm{d}}(\varphi)\rangle}}{|\chi_{\mathrm{L}}|} \simeq -\frac{6}{\pi^2} \frac{\partial^2}{\partial\varphi^2} \sum_{n=1}^\infty \frac{1}{n} \mathcal{S}_n^{(\mathrm{C})}(\omega = 0; \varphi). \tag{5.58}
$$

Disordered Square Quantum Dots

We illustrate the method and present numerical results for the case of square billiards again. For finite magnetic field the integrals over the magnetic vector potential along the paths do not allow for an analytical diagonalization of $\zeta^{(\mathrm{C})}$ as in (5.54) for $H = 0$. However, we can use the fact that all the variations of $\zeta^{(\mathrm{C})}$ occur on classical length scales; rapid oscillations on the scale of λ_F cancel out. This enables an efficient numerical computation. To this end we discretize the configuration space of the square billiard using a lattice

---

[5] It is now assumed that the $\mathcal{S}_n^{(\mathrm{C})}$ in (5.58) contain diagonal terms only.
[6] For diagrammatic approaches to disorder effects on the susceptibility of small diffusive magnetic particles see [292, 293].

with a grid size greater than λ_F. By summing over all trajectories (up to a length $\gg \ell_\phi$) which connect two lattice cells we compute the corresponding matrix elements of $\zeta^{(C)}$ in this representation. After diagonalization we obtain $\overline{\langle \chi^d(H) \rangle}$ from (5.58). This method is not restricted to the square geometry but can in principle be applied to any geometry. The technique, moreover, covers the whole range from the diffusive regime to the clean limit.

We briefly summarize results the of [73] for the magnetic susceptibility at zero field. Figure 5.6 shows $\overline{\langle \chi^d(0) \rangle}$ as a function of l for a typical experimental value of $k_F a = 60$ for the whole range from diffusive to ballistic. The lower and upper curves are for $\ell_\phi/a \approx 10$ and 25, respectively. The susceptibility is large compared to χ_L and always paramagnetic. For $l < a$ there is a linear increase with l. This agrees with previous results for magnetism in diffusive systems [292, 293].

The occurence of the maxima in Fig. 5.6 may be related to the competition between different effects of the impurity scattering on $\zeta^{(C)}$ (5.52): while the single-particle Green functions are exponentially damped with l, l enters as l^{-1} into the prefactor. For larger l, in the regime $a, \ell_\phi < l$, the susceptibility decays exponentially with both l/a and ℓ_ϕ/a.

The disordered-induced magnetic response of the square billiard has previously been studied by Gefen, Braun, and Montambaux [244]. The dashed horizontal line shows their approximate result, predicting a paramagnetic l-independent susceptibility.

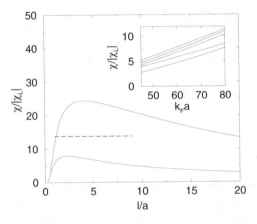

Fig. 5.6. Disorder-induced ensemble-average susceptibility $\overline{\langle \chi^d(0) \rangle}$ for a square geometry as a function of the elastic mean free path l for $k_F a = 60$ and two strengths of inelastic scattering, $\ell_\phi/a \approx 10$ (*lower*, which corresponds to $\gamma/\Delta = 1$) and $\ell_\phi/a \approx 25$ (*upper curve*). The *dashed horizontal curve* indicates the result from [244] for $\gamma/\Delta = 1$. The *inset* shows $\overline{\langle \chi^d(0) \rangle}$ as a function of $k_F a$ for $\ell_\phi/a \approx 10$. From the *top*, the five curves are for values of $l/a = 2, 4, 5, 1,$ and 10. (From [73], by permission)

The inset of Fig. 5.6 shows $\overline{\langle \chi^{\mathrm{d}}(0) \rangle}$ as a function of $k_{\mathrm{F}}a$ for $\ell_\phi/a \approx 10$ for different values of l/a. For all disorder strengths, $\overline{\langle \chi^{\mathrm{d}}(0) \rangle}$ increases linearly with $k_{\mathrm{F}}a$, in line with the results of [244].

Comparison of Semiclassical with Quantum Mechanical Results

We are now in a position to compare the combined semiclassical results for $\overline{\langle \chi^{\mathrm{d}}(H) \rangle}$ and $\overline{\langle \chi^{\mathrm{s}}(H) \rangle}$ with numerical quantum calculations (Appendix A.2). In these calculations temperature smearing, rather than level broadening due to inelastic effects, was introduced.

The full lines in Fig. 5.7 show the numerical quantum results for various disorder strengths as a function of $k_{\mathrm{F}}a$. From the top, the elastic mean free path is $l/a = 8, 4, 2, 1$. The dashed curves in Fig. 5.7 are data taken from the semiclassical evaluation of the contribution of size-induced correlations to the susceptibility, $\overline{\langle \chi^{\mathrm{s}} \rangle}$ (see (5.37) and (5.41)). The contributions of the disorder-induced correlations, $\overline{\langle \chi^{\mathrm{d}} \rangle}$, were computed using the relation $\gamma/\Delta \approx \pi k_{\mathrm{B}}T/\Delta$. The resulting semiclassical evaluation of the total susceptibility, $\overline{\langle \chi^{\mathrm{s}} \rangle} + \overline{\langle \chi^{\mathrm{d}} \rangle}$, is depicted as the dotted curves, showing that the combined semiclassical contributions are indeed close to the quantum results.

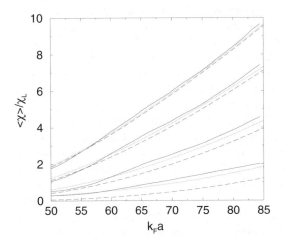

Fig. 5.7. Disorder- and size-induced contributions to the susceptibility. The mean susceptibility is shown for various disorder strengths for small flux $\varphi = 0.15$ and $k_{\mathrm{B}}T/\Delta = 2$ ($L_T/L \approx 2$ at $k_{\mathrm{F}}a = 70$). From the *top*, the elastic MFP is $l/a = 8, 4, 2, 1$ (at $k_{\mathrm{F}}a \approx 70$). The *full* and *dashed curves* are data from a quantum mechanical calculation and from a semiclassical evaluation of the contribution $\overline{\langle \chi^{\mathrm{s}}(H) \rangle}$. The *dotted curves* represent the sum of the semiclassical disorder-induced contribution, $\overline{\langle \chi^{\mathrm{d}}(H) \rangle}$, and the semiclassical data for $\overline{\langle \chi^{\mathrm{s}}(H) \rangle}$. (From [282]; ©1999 by the American Physical Society)

Chaotic Geometries

For systems with a generic chaotic clean counterpart we obtain an analytical estimate for $\overline{\langle \chi^d(0) \rangle}$ from the results for $\mathcal{S}_n^{(C)}$ (see (5.57)). The field-dependent susceptibility has a paramagnetic maximum for zero field [73],

$$\frac{\overline{\langle \chi^d(0) \rangle}}{|\chi_L|} \simeq \frac{96\sigma\ell_\phi^2}{L^4(\ell_\phi + l)} \,. \tag{5.59}$$

A corresponding approximation for $\overline{\langle \chi^s(0) \rangle}$ along the lines of Sect. 4.3.3 (see, e.g., (4.44)) gives, for finite ℓ_ϕ,

$$\frac{\overline{\langle \chi^s(0) \rangle}}{|\chi_L|} \simeq \frac{96\sigma\ell_\phi l}{L^4(\ell_\phi + l)} \,. \tag{5.60}$$

The two contributions (5.59) and (5.60) add up to an *l-independent* magnetic response

$$\frac{\overline{\langle \chi(0) \rangle}}{|\chi_L|} \simeq \frac{96\sigma\ell_\phi}{L^4} \,. \tag{5.61}$$

This is the same as that for clean chaotic systems (see (4.44)) given the assumption of an l-independent variance.

The susceptibility $\overline{\langle \chi^d(0) \rangle}$ as a function of ℓ_ϕ for $l/a = 2$ is shown as the dashed–dotted line in Fig. 5.8. For comparison, the corresponding contribution for the square geometry is also presented in Fig. 5.8 (solid line). The ℓ_ϕ dependence shows a close similarity between the disorder-induced susceptibilities in the square and the chaotic geometry. This may be related to the fact that for trajectories that are (multiply) scattered at impurities the character of the clean geometry, namely regular or chaotic, is of minor importance.

Naturally, the square and the chaotic billiard show quite different behaviors for the size-induced susceptibilities. The short and long dashed lines in Fig. 5.8 display the contributions $\overline{\langle \chi^s(0) \rangle}$ for the chaotic and square geometries, respectively. The results imply that the order-of-magnitude difference in the clean susceptibility according to the shape (chaotic versus integrable) persists in the ballistic regime. For the square a crossover from domination by disorder-induced to domination by size-induced correlations occurs for $\ell_\phi \sim l^2$, which contrasts with chaotic geometries, where the crossover occurs for $\ell_\phi = l$.

Relation to Experiment

Measurements of the orbital magnetism of ballistic systems, which are experimentally realized as semiconductor microstructures, are still rare [57, 220]. In the measurement of the average magnetic susceptibility of an array of ballistic square billiards by Lévy et al. [57], the elastic MFP was estimated to be $l = 4.5$–10 μm, corresponding to a value $l/a = 1$–2. The estimated values

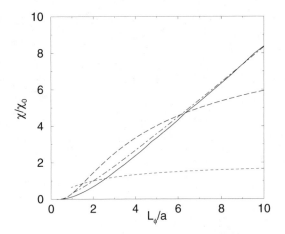

Fig. 5.8. Semiclassical estimates for $\overline{\langle \chi^{\mathrm{d}}(0)\rangle}$ (*dashed–dotted line*) and $\overline{\langle \chi^{\mathrm{s}}(0)\rangle}$ (*short dashed line*) for a generic chaotic geometry and for $\overline{\langle \chi^{\mathrm{d}}(0)\rangle}$ (*solid line*) and $\overline{\langle \chi^{\mathrm{s}}(0)\rangle}$ (*long-dashed line*) for square geometry. The normalization is $\chi_0 = |\chi_{\mathrm{L}}|$ for the square and $|\chi_{\mathrm{L}}|a^3/96\sigma$ for the chaotic geometry ($k_{\mathrm{F}}a = 60$ and $l/a = 2$; from [282]; ©1999 by the American Physical Society)

for the phase coherence length were $\sim (3\text{--}10)a$ and for the thermal cutoff length $L_T/a \sim 2$. Therefore, the length scale ℓ_ϕ (5.53) is determined by the shorter length L_T. Figure 5.7 shows that for these experimental parameters both the disorder- and size-induced correlations are relevant, however, the latter contribution is dominant.

The above remarks hold for white-noise disorder. However, experimental ballistic structures such as those of [57] are usually characterized by smooth disorder potentials. Self-consistent calculations [291] indicate that the characteristic disorder-potential correlation length for the heterostructures of the experiment is of the order of $\xi/a \simeq 0.1$. The effect of smooth disorder on $\overline{\langle \chi^{\mathrm{s}}\rangle}$ has been analyzed in Sect. 5.4.2 (see Fig. 5.3), implying that the reduction of the clean contribution is not as strong as that for white noise disorder and no longer exponential.[7] We therefore expect that in the parameter regime of the experiment the domination of the susceptibility by size-induced correlations is further enhanced when considering smoothed disorder. This leads us to conclude that disorder damping in currently realizable microstructures is sufficiently weak to mask the large effects due to integrability. A more quantitative comparison with experiment, including the consideration of additional Coulomb interaction effects, is performed in the following chapter in Sect. 6.6.

---

[7] Smooth-disorder effects can in principle be incorporated into the calculation of K^{d} by introducing an angle-dependent cross section for the impurity scattering between two successive trajectory segments.

6. Interaction Effects

So far we have studied mesoscopic phenomena on the basis of models of *noninteracting* quasiparticles. Such models have been, in general, successfully applied to a broad variety of observed mesoscopic effects, despite the fact that the Coulomb interaction is neglected.

This concept has been followed for problems in static *charge transport* in particular: prominent properties of both classical and quantum transport in the mesoscopic regime appear to be rather insensitive to the Coulomb interaction. Typical examples are the conductivity in (macroscopic) semiconductor lattices (see Sect. 3.1), conductance fluctuations in disordered systems [2], and transport through ballistic quantum dots (Sect. 3.2.1, [43, 141]). The success of these models may appear surprising since in principle one is dealing with a complicated many-body problem of interacting particles. However, the structures addressed above have in common that they are either extended and thus open, or, in the case of the quantum dots, have rather broad openings at the entrance to the leads allowing for an efficient exchange of charge carriers.

Interaction effects start to play an important role in and have to be considered for time-dependent transport phenomena such as frequency-dependent conductance. This poses the challenging problem of computing the dynamic response of an interacting mesoscopic system.[1]

Interaction effects in static transport gain importance upon reducing the size of the quantum dots and the coupling to the leads, i.e. by changing from open to nearly closed systems which are, for instance, weakly coupled to leads via tunneling barriers. The most striking experimental feature of interactions in this regime is a vanishingly small conductance between pronounced conductance peaks due to the Coulomb blockade [13, 299]. The appearence of nearly equidistant conductance peaks observed in the earlier experiments could be explained by employing a simple charging model in terms of an effective capacitance where the total charging energy merely depends on the total number of electrons in the dot.

Recently, a new generation of experiments on the Coulomb blockade in ballistic and diffusive quantum dots has shown fluctuations in the conductance peaks with respect to both their heights [300] and their spacings [301].

---

[1] For recent work on different aspects of this topic see, for example, [298].

Related effects were reported in the nonequilibrium tunneling conductance of ultrasmall metallic particles [302]. These observations are clear signatures of the interacting many-body character of the quantum dots.

Naturally, the question arises of whether quantum chaos approaches, which have proven very useful in understanding transport for noninteracting particles in open quantum dots, can be generalized to systems with many-body interactions. It appears obvious to employ random-matrix theory (RMT) as one possible approach, since RMT was originally developed and successfully applied to describe the complexity of many-body states in nuclei [23, 303].

Most of the features of the Coulomb blockade peak *height* statistics are indeed well understood within RMT [304]. However, all measured [301] peak *spacing* distributions resemble a Gaussian form, while the simple charging model [299], which properly accounts for the average conductance peak spacing, together with RMT, predicts Wigner–Dyson statistics. Moreover, the observed fluctuations in the ground state energies of interacting quantum dots are found to be considerably larger in most experiments than predicted by RMT. These exciting experiments, which have recently triggered a number of theoretical studies [305, 306], suggest that the widths scale with the charging energy rather than with the mean level spacing.

In addition, RMT approaches have recently been applied to the problem of two interacting electrons, predicting an enhanced two-particle localization length [307]. Such an enhancement has also been found numerically.[2] The theoretical studies have merged into the general topic of interaction and localization effects in disordered systems [311].

The development of semiclassical methods for interaction effects in transport is still in its early stages [312] and thus will not be discussed here.

Instead we focus on the *thermodynamic properties* of isolated systems. There interaction effects are considered to be of special relevance. For example, measurements [210,211,217] of the persistent current of small metal rings have shown an unexpectedly large magnetic response, incompatible with existing theories based on noninteracting particles.[3] This serious disagreement has attracted considerable theoretical activity investigating the role of interaction effects. In particular, the intriguing question was raised of whether the interplay between disorder and interactions in mesoscopic systems [314] may lead to the enhancement of the persistent current. In the next section we shall address this issue on the basis of a semiclassical treatment of interacting systems in the diffusive regime.

The situation is more satisfactory for ballistic microstructures. The semiclassical theory outlined in Sects. 4.5 and 4.6, which is based on the noninteracting particle model, is in line with the related experiments [57, 220] for square and ring geometries with regard to the magnitude of the magnetic

---

[2] See [308], [309] and the comment by [310] for a discussion of this issue.

[3] For recent reviews see [8, 218, 219].

response at low temperature. However, the experimental temperature behavior, in particular the nonvanishing susceptibility of the square quantum dots observed at rather high T, cannot be accounted for in any known model using noninteracting particles. This is one motivation for a study of interaction effects on orbital magnetism. Furthermore, the known relevance of Coulomb interaction to the magnetism of disordered systems calls for an investigation in the ballistic regime, too.

In addition, we wish to generalize the investigation of quantum chaos, which has been predominantly addressed for noninteracting particles, to interacting quantum systems. In particular, we shall consider the role of the classical dynamics of the *noninteracting* system in the quantum properties of its *interacting* counterpart. To this end, we assume a screened interaction and use a corresponding quantum mechanical many-body perturbation theory as a starting point. A semiclassical evaluation of the relevant diagrams of the perturbation series for the thermodynamic potential allows one to express the latter in terms of an essentially classical operator for the underlying noninteracting system. We pursue this approach for problems with both diffusive and ballistic noninteracting dynamics. By making connection with the classical dynamics the semiclassical approach provides an intuitive physical picture of the interplay between the interaction and the disorder in diffusive systems and between the interaction and the confinement potential in ballistic quantum dots. As one main result, we show in the latter case that, intringuingly, the thermodynamic properties scale differently with the Fermi energy for interacting systems with chaotic and integrable counterparts. This difference, which is a correlation effect, stems from the semiclassical off-diagonal path contributions present in regular systems.

To make contact with the preceding two chapters we shall review results for the orbital magnetic response of ensembles of diffusive and ballistic systems which were first obtained in [74, 75, 313]. In the next two sections we semiclassically evaluate the relevant interaction diagrams. The spirit of this approach is similar to that used to compute disorder-induced spectral correlations in the preceding chapter.

In Sect. 6.3 we compute the interaction contribution to the persistent current of metal rings and the susceptibility of singly connected two-dimensional diffusive systems such as disordered quantum dots. We precisely recover some results from quantum diagrammatic perturbation theory, showing that the semiclassical approach is on the same level of approximation.

In Sect. 6.4 we study the susceptibility of integrable and chaotic ballistic quantum dots. We again refer to the ensemble of square billiards as an example for numerical calculations. In Sect. 6.5 we compare the interaction contributions for chaotic and regular systems and close with a comparison with experiment.

6.1 Diagrammatic Perturbation Theory

We aim at the interaction-induced orbital magnetism of diffusive or ballistic mesoscopic quantum systems. As we shall see below, the contribution to the grand canonical potential due to electron–electron interactions, which we denote by Ω^{ee}, does not average out under ensemble averaging. Therefore, contrary to the noninteracting case (Sect. 4.2.3), we need not consider here canonical corrections, which merely give higher-order $\hbar$ corrections.

As discussed in Sect. 4.2.3, the magnetic moment of a ring-type structure threaded by a flux $\phi = HA$ (where A is the enclosed area) is usually described by the related persistent current I, while the magnetic response of a singly-connected quantum dot will be measured in terms of its susceptibility χ. The interaction contributions to both are then given by (A denoting the area of the sample)

$$I^{\mathrm{ee}} \equiv -c\frac{\partial \Omega^{\mathrm{ee}}}{\partial \phi} \quad ; \quad \chi^{\mathrm{ee}} \equiv -\frac{1}{A}\frac{\partial^2 \Omega^{\mathrm{ee}}}{\partial H^2} \; . \tag{6.1}$$

To calculate Ω^{ee} we start from the high-density expansion (random-phase approximation (RPA)) of the thermodynamic potential. To obtain the interaction contribution to the magnetic response one has to extend the RPA series by including additional interaction corrections from diagrams of the Cooper channel [281]. This was originally pointed out and performed in the context of superconducting fluctuations. The same procedure was later applied to disordered normal metals [315–319]. The relevant series of Cooper-like terms is shown in Fig. 6.1. If these diagrams are properly resummed, such expansions usually yield reliable results even beyond the high-density limit. This allows us to use this approach not only for disordered metal systems but also for (semiconductor) quantum dots where the parameter $r_{\mathrm{s}} = r_0/a_0$, which is small at high densities, is about 2. Here, πr_0^2 is the average area per electron, and a_0 is the effective Bohr radius in the material.

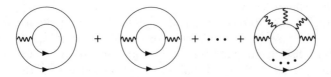

Fig. 6.1. Leading Cooper channel diagrams for the interaction contribution to the thermodynamic potential

A high-density approximation amounts to the use of a screened Coulomb interaction U (wavy lines in Fig. 6.1), which we shall treat as local:

$$U(\boldsymbol{r} - \boldsymbol{r}') = \lambda_0 N(0)^{-1}\delta(\boldsymbol{r} - \boldsymbol{r}'). \tag{6.2}$$

Here, $N(0)$ denotes the density of states per unit volume and the book-keeping index $\lambda_0 = 1$ identifies the order of perturbation. The replacement of the screened interaction by a local one can be justified for (disordered) metal rings,[4] but may appear questionable for two-dimensional semiconductor quantum dots, where screening is supposed to be weaker. However, as we shall show in Sect. 6.4, the local approximation is justified for the calculation of the averaged magnetic response. For the local interaction, the direct and exchange terms differ only by a factor of minus two (owing to the spin summation and an extra minus sign in the exchange term). Then, the perturbation expansion for Ω^{ee} can be formally expressed as [316, 317]

$$\Omega^{ee} = -\frac{1}{\beta} \sum_{n=1}^{\infty} \frac{(-\lambda_0)^n}{n} \sum_{\omega} \int d\boldsymbol{r}_1 \dots d\boldsymbol{r}_n \, \Sigma_{\boldsymbol{r}_1, \boldsymbol{r}_2}(\omega) \dots \Sigma_{\boldsymbol{r}_n, \boldsymbol{r}_1}(\omega) \quad (6.3a)$$

$$= \frac{1}{\beta} \sum_{\omega} \text{Tr} \left\{ \ln[1 + \lambda_0 \hat{\Sigma}(\omega)] \right\} . \quad (6.3b)$$

The sum is taken over the (bosonic) Matsubara frequencies $\omega \sim \omega_m = 2m\pi/\beta$. The particle–particle propagator $\hat{\Sigma}(\omega)$ is given in terms of products of finite-temperature Green functions [281]

$$\Sigma_{\boldsymbol{r}, \boldsymbol{r}'}(\omega) = \frac{1}{\beta N(0)} \sum_{\epsilon_n}^{\mu} \mathcal{G}_{\boldsymbol{r}, \boldsymbol{r}'}(\epsilon_n) \mathcal{G}_{\boldsymbol{r}, \boldsymbol{r}'}(\omega - \epsilon_n) . \quad (6.4)$$

Here, the sum runs over the (fermionic) Matsubara frequencies $\epsilon_n = (2n + 1)\pi/\beta$. The frequency sum is cut off at the Fermi energy μ [317], which takes care of the fact that the short-length (high-frequency) behavior is already implicitly included in the screened local interaction.

The finite-temperature Green functions shown in Fig. 6.1 are given in terms of retarded and advanced Green functions $G^{\pm}$ (of the noninteracting system):

$$\mathcal{G}_{\boldsymbol{r}, \boldsymbol{r}'}(\epsilon_n) = \Theta(\epsilon_n) G^+(\boldsymbol{r}, \boldsymbol{r}'; \mu + i\epsilon_n) + \Theta(-\epsilon_n) G^-(\boldsymbol{r}, \boldsymbol{r}'; \mu + i\epsilon_n) . \quad (6.5)$$

For ballistic systems they have to obey the boundary conditions of the confining potential; for diffusive systems they include the presence of a disorder potential.

6.2 Semiclassical Formalism

Following [74, 75] we now compute $\Sigma_{\boldsymbol{r}, \boldsymbol{r}'}(\omega)$ semiclassically. Accordingly, we assume that the Fermi wave vector k_F is small compared to the elastic mean free path l in the diffusive case or the system size a for a ballistic quantum dot. Furthermore, the magnetic field is assumed to be classically weak, i.e. the cyclotron radius $r_{cyc} \gg l$ or $r_{cyc} \gg a$, respectively.

---

[4] For a discussion of short-range interactions see e.g. [317, 319] and Sect. B of [218].

As usual we employ the semiclassical approximation (2.3) for the retarded Green function, given as the sum of the contributions

$$G^{+;j}(\boldsymbol{r},\boldsymbol{r}';E) \simeq \tilde{D}_j \exp\left(\mathrm{i}S_j/\hbar - \mathrm{i}\pi\eta_j/2\right) \tag{6.6}$$

of all classical paths j from $\boldsymbol{r}$ to $\boldsymbol{r}'$. The semiclassical representation allows one to isolate the temperature-dependent (via the Matsubara frequencies) and H-field-dependent parts in the finite-temperature Green function: employing $\partial S_j/\partial E \simeq t_j$ (2.16) and $\partial S_j/\partial H \simeq (e/c)A_j$ (2.15), where t_j and A_j are the traversal time and area, we have [75]

$$G^{+;j}(\boldsymbol{r},\boldsymbol{r}';\mu+\mathrm{i}\epsilon_n,H) \tag{6.7}$$

$$\simeq G^{+;j}(\boldsymbol{r},\boldsymbol{r}';\mu,H{=}0) \times \exp\left(-\frac{\epsilon_n t_j}{\hbar}\right) \times \exp\left(\frac{\mathrm{i}2\pi H A_j}{\phi_0}\right).$$

As usual, temperature gives rise to an exponential suppression of long paths. The semiclassical form of G^- is obtained by using the relation $G^-(\boldsymbol{r},\boldsymbol{r}';E) = [G^+(\boldsymbol{r}',\boldsymbol{r};E^*)]^*$.

$\Sigma_{\boldsymbol{r},\boldsymbol{r}'}(\omega)$ in (6.4) represents semiclassically a sum over pairs of trajectories connecting $\boldsymbol{r}$ and $\boldsymbol{r}'$. Contributions from products G^+G^+ and G^-G^- as well as off-diagonal pairs (of different paths) in G^+G^- usually contain highly oscillating contributions. They do not survive an ensemble (disorder) average and can be neglected.[5] In contrast, in the diagonal terms of G^+G^-, composed of a path j and its time reverse, the sum of the dynamical phases $\exp[\mathrm{i}S_j(H{=}0)/\hbar]$ cancels while retaining a magnetic-field dependence $\exp(4\pi\mathrm{i}HA_j)$. Thus, terms involving such trajectory pairs persist upon averaging. The behavior under averaging is one criterion for distinguishing semiclassically relevant from irrelevant diagrams in the perturbative expansion of I^{ee} and χ^{ee} (via Ω^{ee}). Furthermore, the relevant terms must be H-field-sensitive and of leading order in $\hbar \sim 1/k_{\mathrm{F}}l$ or $1/k_{\mathrm{F}}a$, respectively. A more detailed analysis shows [75,313] that only the Cooper series shown in Fig. 6.1 obeys all these conditions. In particular, all terms in the Cooper series are of the same order of $\hbar$. Thus, the entire series has to be considered.

Combining (6.5), (6.6), and (6.7) with (6.4) yields for the diagonal part of $\Sigma_{\boldsymbol{r},\boldsymbol{r}'}$, after performing the Matsubara sum [74,75],

$$\Sigma_{\boldsymbol{r},\boldsymbol{r}'}^{\mathrm{D}}(\omega) \simeq \frac{\hbar}{\pi N(0)} \sum_{j:\boldsymbol{r}\to\boldsymbol{r}'}^{L_j>\lambda_{\mathrm{F}}/\pi} |\tilde{D}_j|^2 \frac{R\left(2t_j/\tau_T\right)}{t_j} \exp\left(\frac{\mathrm{i}4\pi HA_j}{\phi_0}\right) \exp\left(-\frac{\omega t_j}{\hbar}\right). \tag{6.8}$$

The sum runs over all trajectories longer than λ_{F}/π (corresponding to the upper bound μ in the Matsubara sum in (6.4)). The temperature dependence in (6.8) enters through the familiar function $R(x) = x/\sinh(x)$, which introduces the timescale (see (2.50))

$$\tau_T = \frac{\hbar\beta}{\pi} \tag{6.9}$$

---

[5] There are exceptional cases (integrable ballistic systems) where off-diagonal contributions must be considered (as will be shown in Sect. 6.4).

and (for billiard systems) the thermal length scale $L_T = \tau_T v_F$.

As a result of (6.3a), Ω^{ee} is given as a sum over trajectory pairs which visit successively the points r_j, representing local interaction events, before returning to the initial point. This is illustrated in Fig. 6.2 for the case of a clean billiard. Typical pairs of equal paths are shown, contributing to the diagonal Cooper channel to first and fourth order in the interaction. The combined effect of a local interaction (at r) and the boundary already allows for flux-enclosing paths to first order. In contrast, flux can be enclosed in the bulk only with two and more interactions. We stress the following:

(i) The semiclassical framework enables us to reduce the original fully inter-acting *quantum* problem to the evaluation of the quantity Σ^D, which no longer exhibits oscillations on the scale of λ_F but only varies on *classical* length scales. This allows for an efficient calculation.

(ii) The semiclassical representation (6.8) of $\hat{\Sigma}$ is rather general since we have not yet made any assumption about the character of the underlying classi-cal mechanics of the system. Therefore it applies, as it stands, to diffusive as well as ballistic structures, and in the latter case irrespective of whether the system is integrable or chaotic. This is related to the fact that Σ^D represents the "classical" part (apart from the flux-dependent phase) of the averaged particle–particle propagator. Trace integrals, therefore, do not involve stationary-phase integrations as for objects containing quan-tum oscillations, which would yield a different $\hbar$ dependence according to the character of the classical dynamics.

a)   b)

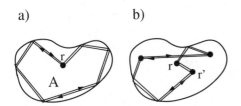

Fig. 6.2. Sketch of (diagonal) pairs of equal flux-enclosing paths in a ballistic quantum dot contributing to Ω^{ee} in first order (**a**) and fourth order (**b**) in the interaction

On the basis of (6.8) we discuss two important applications, the orbital magnetism of diffusive and ballistic devices. The character of the classical dynamics enters via the amplitudes $\tilde{D}_j$.

6.3 Orbital Magnetism of Interacting Diffusive Systems

For disordered systems it proves convenient to express Σ^D in terms of classical probabilities for diffusive motion. For that purpose we introduce an additional time integration in (6.8) in order to make use of the relation (2.7b), which relates the weights $|\tilde{D}_j|^2$ to the classical probability for propagation from r to r' in time t. An nth order (diagonal) contribution to Ω^{ee} in (6.3a) contains the joint conditional return probability $P(r_1, \ldots, r_n, r_1; t_1, \ldots, t_n | A)$ for successively visiting the points r_i and enclosing an area A (t_i denotes the time between r_i and r_{i+1}). For diffusive motion the probability is multiplicative, namely

$$\int dr_1 \ldots dr_n P(r_1, \ldots, r_n, r_1; t_1, \ldots, t_n | A) = \int dr \, P(r, r; t_{tot} | A) \quad (6.10)$$

with $t_{tot} = \sum t_i$. Upon using (2.7b) and (6.10) in (6.8), the diagonal contribution Σ^D to Ω^{ee} then yields for diffusive systems [74]

$$\Omega^{ee,D} = \sum_n \Omega_n^{ee,D} = \frac{1}{\beta} \int dr \int dt \coth\left(\frac{t}{\tau_T}\right) K(t) \, \mathcal{A}(r, t; H) . \quad (6.11)$$

The $\coth(t/\tau_T)$ arises from the ω sum in (6.3a), and

$$K(t) \equiv \sum_n K_n(t) \quad (6.12)$$

$$= -\sum_n \frac{(-\lambda_0)^n}{n} \left\{ \int \prod_{i=1}^n \left[\frac{dt_i R(2t_i/\tau_T)}{t_i} \right] \delta(t - t_{tot}) \right\} ,$$

$$\mathcal{A}(r, t; H) \equiv \int dA \cos\left(\frac{4\pi H A}{\phi_0}\right) P(r, r; t | A) . \quad (6.13)$$

The kernel $K(t)$ contains system-independent temperature effects, while $\mathcal{A}$ takes into account the system-specific classical return probability and the H-field dependence. Equations (6.11)–(6.13) represent a convenient starting point for computing the orbital response of disordered solids.

6.3.1 Disordered Rings

Consider a thin disordered ring of width b, radius r, and circumference a. If $a \gg l, b$ we can use the one-dimensional diffusion approximation for the motion of particles along the ring. $D = v_F l / d$ is the diffusion constant (in d dimensions). The area enclosed is given in terms of the winding number k of paths encompassing the magnetic flux. Therefore

$$P(r, r; t | A) = \frac{1}{\sigma} \frac{1}{\sqrt{4\pi D t}} \sum_{k=-\infty}^{+\infty} \exp\left(-\frac{k^2 a^2}{4Dt}\right) \delta(A - \pi k r^2) , \quad (6.14)$$

where σ is the cross section of the ring. Owing to the disorder average the classical return probability does not depend on r.

We first compute the first-order interaction contribution, $\Omega_1^{ee,D}$, to illustrate the main ideas. From (6.12) we have

$$K_1(t) = \lambda_0 \frac{R(2t/\tau_T)}{2t} . \tag{6.15}$$

After combining $K_1(t)$ with the coth function in (6.11) we find

$$\Omega_1^{\mathrm{ee,D}} = \lambda_0 \frac{a\hbar}{\pi} \sum_{k=-\infty}^{+\infty} \cos\left(\frac{4\pi k\phi}{\phi_0}\right) g_k(T), \tag{6.16}$$

with the temperature-dependent function

$$g_k(T) = \int_0^\infty dt \frac{R^2(t/\tau_T) \exp\left[-(ka)^2/(4Dt)\right]}{t^2 \sqrt{4\pi Dt}} . \tag{6.17}$$

The flux derivative (6.1) then gives the first-order interaction contribution to the persistent current,

$$I_1^{\mathrm{ee}} = \lambda_0 \frac{ae}{\pi} \sum_{k=-\infty}^{+\infty} k \sin\left(\frac{4\pi k\phi}{\phi_0}\right) g_k(T). \tag{6.18}$$

This first-order result was first obtained by Ambegaokar and Eckern [318] using quantum diagrammatic techniques. It has also been derived semiclassically by Montambaux [320]. However, as pointed out above, the higher-order terms of the Cooper series (Fig. 6.1) are of the same order in $\hbar$ and must be considered to get reliable quantitative results. Indeed, higher-order diagrams lead to a renormalization of the coupling constant [315, 317, 319].

To show this semiclassically we have to incorporate the full kernel $K(t)$ into (6.11). Without going into the technical details we note that the full kernel in (6.12) can be expressed in terms of $K_1(t)$:[6]

$$K(t) \simeq \frac{2}{\lambda_0 \ln(k_{\mathrm{F}} L^*)} K_1(t) \tag{6.19}$$

with $L^* = \min(v_{\mathrm{F}} t, L_T/4)$. Therefore, the common effect of all higher-order terms can be considered as a renormalization of the original coupling constant $\lambda_0 \equiv 1$ (for $K_1(t)$, see (6.15)) to $2/\ln(k_{\mathrm{F}} L^*)$.

Hence, the persistent current from the entire interaction contribution is reduced accordingly, and reads [74]

$$I^{\mathrm{ee}} = \frac{2ae}{\pi \ln(k_{\mathrm{F}} L^*)} \sum_{k=-\infty}^{+\infty} k \sin\left(\frac{4\pi k\phi}{\phi_0}\right) g_k(T). \tag{6.20}$$

For rings the length scale $v_{\mathrm{F}} t$ is effectively given by the average length $L_k = v_{\mathrm{F}}(ka)^2/4D$ of diffusive trajectories with winding number k. For low temperature ($L_T \gg L_k$) $\lambda_0 \equiv 1$ is replaced by $2/\ln(k_{\mathrm{F}} L_T/4)$. At higher temperature ($L_T \ll L_k$) the coupling constant is renormalized to $1/\ln(k_{\mathrm{F}} L_k)$. The result (6.20) for the renormalized persistent current (including the full Cooper series) is equivalent (for both limits discussed above) to a previous

---

[6] See [313]. Equation (6.19) is valid if $\ln(k_{\mathrm{F}} L^*) \gg 1$, which holds true in the diffusive regime for $\ln(k_{\mathrm{F}} l) \gg 1$.

result by Eckern [319]. This was obtained using quantum perturbation theory on the basis of the same Cooper series as above. As was pointed out in [318], the temperature dependence of the kth harmonic is well approximated by $\exp(-k^2 k_\mathrm{B} T/3 E_\mathrm{erg})$ in the regime $k^2 k_\mathrm{B} T < E_\mathrm{erg}$ (where $E_\mathrm{erg} = \hbar/t_\mathrm{erg}$ is the Thouless energy).

6.3.2 Relation to Experiments and Other Theoretical Approaches

Measurements of the persistent current of small metal rings have shown an unexpectedly large magnetic response and have motivated a considerable number of theoretical approaches in the past. In the first generation of experiments, Chandrasekhar et al. [211] studied the magnetic response of a single Au ring, while Lévy et al. [210] observed an average persistent current in an array of 10^7 Cu rings. A more recent measurement by Mohanty et al. [217] addresses the magnetic response of a small array of 30 isolated gold loops. All experimental samples operate in the diffusive regime. In the latter experiment, for instance, the rings have a diameter of 2.6 µm and a thickness of 60 nm, while the elastic mean free path is $l \simeq 100$ nm. Thus the assumption in the proceeding section of a thin ring with effectively one-dimensional diffusive dynamics applies well.

One class of theoretical approaches uses models of noninteracting particles in the diffusive regime. They are based partly on numerical calculations [213, 214, 216], on diagrammatic Green function methods [216], on semiclassical approximations [76] (as briefly discussed in Sect. 4.3.4), or on supersymmetry techniques [8, 321], to name a few. Typically, these approaches give results nearly two orders of magnitude smaller than the measured persistent currents.[7]

This discrepancy has pointed towards the importance of interaction effects. The results of the early work by Ambegaokar and Eckern [318, 319], which were derived semiclassically above, give an amplitude of the first-order average persistent current which is on the order of the measured value. However, the renormalization of the coupling constant reduces the full magnetic response by a factor of ~ 5. We note that the functional form of the temperature dependence (exponential T damping [319]) is in line with recent experimental results [217] although the exponent is a factor of ~ 3 off.

Here, we do not intend to systematically review the numerous other theoretical approaches based on interaction effects (for recent reviews see [8, 218, 219]), but name only a few examples: in a comprehensive and detailed work Müller-Groeling and Weidenmüller [218] studied the persistent current in one- and two-dimensional rings, going beyond Hartree–Fock approximation. They showed that the interaction counteracts the suppression of the

---

[7] In contrast, experimental results for the persistent current in the ballistic regime [220] are of the same order as those of our noninteracting theory (see Sect. 4.6.1).

persistent current by impurities. Several numerical studies, based on tight-binding models, consider the one-dimensional case for both short-ranged [322] and long-ranged (e.g. [323]) interactions. There also exist studies based on continuous models [324] and numerical approaches including three dimensions [325].

However, the approaches cited above yield different results depending on the method used, the number of dimensions considered, the strength of the disorder and the interaction potential, and the inclusion of spin. Therefore, to my knowledge, a conclusive, precise calculation, which would account for all experimental conditions of possible relevance, is still missing. This would presumably amount to an extension beyond the Hartree–Fock approximation and would require the consideration of correlations (for more than one dimension), spin effects, and finite-temperature effects. Despite the intense theoretical efforts the observed magnitude of the persistent current is not yet understood and remains as one of the open questions in mesoscopic physics.

6.3.3 Two-Dimensional Diffusive Structures

The exponential temperature dependence of I^{ee} in rings can be traced back to the fact that the shortest flux-enclosing orbits have a finite minimum length given by the circumference. In singly connected systems such as the disordered bulk and quantum dots, there exists no such finite length scale imposed by the geometry. Thus, a different temperature behavior is expected.

We consider in the following a two-dimensional singly connected diffusive quantum dot. To calculate the thermodynamic potential we follow [74] and employ the general renormalization property (6.19) for diffusive systems. Hence, $\Omega^{ee,D}$ for the entire Cooper series (6.11) can be represented as

$$\Omega^{ee,D} = \frac{1}{\beta} \int d\boldsymbol{r} \int dt \, \frac{1}{\ln(k_F v_F t)} \frac{\tau_T}{t^2} \, R^2\left(\frac{t}{\tau_T}\right) \mathcal{A}(\boldsymbol{r}, t; H) . \tag{6.21}$$

Here we have used the case $L^* = v_F t$ in (6.19) since the R^2 function causes the main contribution to the integral to come from times smaller than τ_T. The conditional return probability (4.47) for a given enclosed area in two dimensions, which enters into $\mathcal{A}$, is conveniently expressed in terms of its Fourier transform:

$$P(\boldsymbol{r}, \boldsymbol{r}; t|A) = \frac{1}{8\pi^2} \int dk \, |k| \, \frac{\exp(ikA)}{\sinh(|k|Dt)} . \tag{6.22}$$

Using this in (6.13) we find

$$\mathcal{A}(\boldsymbol{r}, t; H) = \frac{1}{4\pi D} \frac{R(t/\tau_H)}{t} , \tag{6.23}$$

where the function R is of different origin than in (6.21). In the last equality we have introduced the magnetic time

$$\tau_H = \frac{\phi_0}{4\pi HD} = \frac{L_H^2}{2D} \ . \tag{6.24}$$

This can be expressed through the square of the magnetic length L_H^2 (see (5.25)), which is related to the area enclosing one flux quantum at a given field strength H.

After introducing the expression (6.23) into (6.21) and taking the second derivative with respect to the field, we find for the average susceptibility in two dimensions (normalized to the Landau susceptibility of noninteracting particles)

$$\frac{\chi^{\rm ee}}{|\chi_{\rm L}|} = -\frac{12}{\pi}(k_{\rm F}l) \int_\tau^\infty \frac{dt}{t \ln(k_{\rm F} v_{\rm F} t)} \ R^2 \left(\frac{t}{\tau_T}\right) R'' \left(\frac{t}{\tau_H}\right) \ . \tag{6.25}$$

Here, R'' denotes the second derivative of R and we have used $D = v_{\rm F}l/2$.

In the above time integrals the elastic scattering time $\tau = l/v_{\rm F}$ is introduced as a lower bound since for backscattered paths with times shorter than τ the diffusion approximation (6.22) no longer holds. Flux-enclosing short paths with $t < \tau$, which arise from (higher-order) interaction events, contribute to the clean bulk magnetic response. This is, however, much smaller than the disorder-mediated interaction contribution [326].

Furthermore, (6.25) holds true for a quantum dot of finite size a only as long as the upper cutoff time $t^* \equiv \min(\tau_T, \tau_H)$ is smaller than the ergodic or Thouless time $t_{\rm erg} = a^2/D$ (see (5.43)). For times larger than $t_{\rm erg}$ the classical dynamics begins to behave ergodically and the diffusion approximation is no longer valid.

For $\tau \ll t^* < t_{\rm erg}$ the integral in (6.25) can be evaluated approximately: by replacing $R(t/\tau_T)$ and $R''(t/\tau_H)$ by $R(0) = 1$ and $R''(0) = -1/3$, respectively, and introducing the upper cutoff t^*, the remaining integral gives

$$\int_\tau^{t^*} \frac{dt}{t \ln(k_{\rm F} v_{\rm F} t)} = \ln \left\{ \frac{\ln[k_{\rm F} \, v_{\rm F} \min(\tau_T, \tau_H)]}{\ln(k_{\rm F}l)} \right\} \ . \tag{6.26}$$

The log–log behavior related to the $1/(t \ln t)$ form of the integrand reflects the wide length and area distribution of contributing paths, with lengths ranging from about $v_{\rm F}\tau$ up to $v_{\rm F}t^*$. In contrast, in the ring geometry discussed in the previous section the temperature dependence is exponential because the minimum length of flux-enclosing trajectories is the circumference.

The interaction-induced averaged susceptibility of a diffusive two-dimensional structure then reads [74]

$$\frac{\chi^{\rm ee}}{|\chi_{\rm L}|} \simeq \frac{4}{\pi}(k_{\rm F}l) \ln \left\{ \frac{\ln[k_{\rm F} \, v_{\rm F} \min(\tau_T, \tau_H)]}{\ln(k_{\rm F}l)} \right\} \ . \tag{6.27}$$

As a consequence of (6.9) and (6.24) one thus finds a log–log T dependence for $\tau_T < \tau_H$ and a log-log H dependence for $\tau_T > \tau_H$. Furthermore, the interaction contribution to the susceptibility of a diffusive quantum dot or the (phase-coherent) bulk is *paramagnetic* and is enhanced in magnitude by a factor $k_{\rm F}l$ compared to the clean Landau susceptibility $\chi_{\rm L}$.

Equation (6.27) agrees with results from Aslamazov and Larkin [315] and Altshuler, Aronov, and Zyuzin [316, 317] obtained with quantum diagrammatic perturbation theory. This agreement, which we also found in the previous section for the persistent current of ring structures, points towards the equivalence of semiclassical and quantum mechanical approaches for the evaluation of interaction diagrams in diffusive systems. This observation may be traced back to the fact that "quantum" diagrammatic perturbation theory relies on the existence of the small parameter $1/k_F l$ which can be viewed as already being a semiclassical approximation.

6.4 Orbital Magnetism of Interacting Ballistic Quantum Dots

In the previous section we showed how disorder enhances the interaction contribution to the orbital magnetic response. Impurity scattering provides a mechanism for backscattering and even first-order interaction diagrams contain flux-enclosing returning orbits, thus contributing to the magnetism. In contrast, in the clean bulk only terms in the Cooper series which include three or more interaction events allow for flux-enclosing returning paths [326].

In ballistic quantum dots, however, the confinement already allows for closed trajectories in the first order of the interaction. This will modify the clean-bulk value considerably. Following [75], we present here such effects arising from the interplay between the confining and interaction potentials. The usual techniques to deal with Coulomb interactions in solids, which rely on translational invariance by working in momentum space, are not applicable to finite-size systems. Hence, we use again semiclassical techniques in configuration space, similar to those used in Sect. 5.5 to compute disorder-induced spectral correlations.

To be specific, we investigate the magnetic response of an ensemble of ballistic quantum dots. To this end we consider the contribution from the Cooper series (Fig. 6.1) which provides the leading-order diagrams relevant to the orbital magnetism as discussed in Sect. 6.2. There we derived the general semiclassical approximation (6.8) for the diagonal part of the particle–particle propagator. As already stressed, Σ^D is an essentially classical operator and applies to any type of classical motion. The diagonal approximation $\Omega^{ee,D}$ of the thermodynamic potential is obtained by using Σ^D in the perturbation expansion (6.3a). A calculation of the trace integrals appearing in it has usually to be performed numerically since for ballistic dynamics we cannot rely on properties such as (6.10) in the diffusive case.

Since Σ^D contains no rapidly oscillating phases, trace integrals are performed in the same way for both regular and chaotic dynamics and therefore both types of classical behavior lead to the same $\hbar$ dependence. Thus, thermodynamic properties of integrable and chaotic systems based on the diagonal part of Σ exhibit the same parametric scaling with the Fermi energy. This interaction contribution and the issue of renormalization in the ballistic regime due to higher-order terms will be considered in this section.

However, in ensembles of systems with regular motion, additional *off-diagonal* contributions to Σ persist which do *not* average out. These terms will be discussed in Sect. 6.4.4. Their first-order interaction diagram is dominant and therefore these terms are not renormalizable: a subtle property which leads to a parametric difference between interacting systems with noninteracting chaotic and regular behaviors.

a)   b)   c)

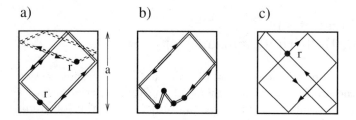

Fig. 6.3. Typical pairs of flux-enclosing paths in a square contributing to the average susceptibility. **(a)**, **(b)** Pairs of equal trajectories belonging to the diagonal channel to first and fourth order in the interaction. An orbit pair entering into the nondiagonal channel is shown in **(c)**

6.4.1 Ensemble of Squares

In the following we consider again an ensemble of square quantum dots as representing regular systems. This working example allows one to illustrate the main physical effects and to make contact with experiment and the calculations for the noninteracting case.

For the square, the classical amplitudes $|\tilde{D}_j|^2$, given by (4.59), merely depend on the path lengths. Then the diagonal part of the particle–particle operator reads

$$\Sigma^{\mathrm{D}}_{\boldsymbol{r},\boldsymbol{r}'}(\omega) \simeq \frac{1}{\pi} \sum_{j:\boldsymbol{r}\to\boldsymbol{r}'}^{L_j>\lambda_{\mathrm{F}}/\pi} \frac{R(2L_j/L_T)}{L_j^2} \exp\left(\frac{\mathrm{i}4\pi H A_j}{\phi_0}\right) \exp\left(-\frac{\omega t_j}{\hbar}\right). \quad (6.28)$$

All paths connecting $\boldsymbol{r}$ with $\boldsymbol{r}'$ are most easily obtained by the method of images, as discussed in Sect. 4.5.

Even for the simple geometry of a square billiard, the calculation of the integrals $A_j(\boldsymbol{r},\boldsymbol{r}')$ of the magnetic vector potential along the paths does not allow for an analytical treatment of the integrals (6.3a) leading to $\Omega^{\mathrm{ee,D}}(H)$. However, all variations of Σ^{D} happen only on classical scales since oscillations on the scale of the Fermi wavelength are already averaged out. This allows for an efficient numerical computation. To this end, we discretize the configuration space of the square billiard and calculate the matrix elements of Σ^D in the basis set representing the discretization. The susceptibility χ^{ee} is then obtained either by first taking analytically the flux derivatives in (6.3b), followed by matrix operations, or by first diagonalizing Σ^D and then taking the H derivative of [75]:

$$\Omega^{ee,D}(H) = -\frac{1}{\beta} \sum_{n=1}^{\infty} \frac{(-\lambda_0)^n}{n} \sum_{\omega} \text{Tr}[\Sigma^D(\omega, H)]^n$$

$$= -\frac{1}{\beta} \sum_{n=1}^{\infty} \frac{(-\lambda_0)^n}{n} \sum_{k=1}^{N^2} [\sigma_k(\omega, H)]^n. \tag{6.29}$$

Here, σ_k denotes the eigenvalues of Σ^D for an $N \times N$ grid. For squares of size a the length a/N of a cell used in the numerical calculations should be small compared to a and L_H, but may be larger than λ_F, since we are essentially dealing with a classical operator.

6.4.2 First-Order Diagonal Channel

We begin with the treatment of the diagonal interaction contribution to χ^{ee} to first order in the coupling constant. As sketched in Fig. 6.3a, two types of shortest flux-enclosing trajectories exist starting and ending at a single interaction event, namely the usual periodic orbits of the family $(1,1)$ and parallelogram orbits, which may be labeled by $(1,1/2)$. At low temperature higher repetitions as well as closed orbits of other topology also contribute to the magnetic response. The numerically calculated susceptibility contribution $\chi_1^{ee,D}(H=0)$ from the $n=1$ term of the sum in (6.29) is depicted in Fig. 6.4 (full lines) as a function of $k_F a$ for different temperatures. It exceeds the average susceptibility χ_{11} from the noninteracting model (dashed, see (4.75)) by more than a factor of 2. The dependence of $\chi_1^{ee,D}$ on temperature and Fermi energy is the same as for χ_{11}. In particular, the first-order susceptibility is exponentially suppressed at high T (for $L_T \leq a$).

Before discussing the effect of higher-order interaction contributions we employ the first-order calculations to study the validity of the model (6.2) of a local interaction. To this end we compare the results from this model with results where the δ-like interaction potential is replaced by the Thomas–Fermi potential of the screened interaction. In two dimensions this reads, in momentum representation,

$$\hat{U}(k) = \frac{2\pi e^2}{k + k_{\text{TF}}}, \tag{6.30}$$

where $k_{\text{TF}} = \sqrt{2} r_s k_F = 2\pi e^2 N(0)$ denotes the Thomas–Fermi screening wave vector. A more detailed semiclassical analysis of the first-order Cooper diagram for a nonlocal potential shows that the use of the screened interaction amounts to replacing the coupling constant $\lambda_0 = 1$ by the difference between the (paramagnetic) Hartree and Fock contributions,

$$\frac{\lambda_0}{N(0)} \quad \longrightarrow \quad 2\hat{U}(|\boldsymbol{k}_i - \boldsymbol{k}_f|) - 1\hat{U}(|\boldsymbol{k}_i + \boldsymbol{k}_f|) \,. \tag{6.31}$$

Here, $\boldsymbol{k}_i$ and $\boldsymbol{k}_f$ are the initial and final wave vectors of a given closed trajectory at the center of the interaction potential. The first-order result for the susceptibility based on the screened interaction is shown as dashed–dotted lines in Fig. 6.4. The difference on the order of 10% shows that the use of a δ-like interaction, which amounts to neglecting the momentum difference of the Thomas–Fermi potential, is justified.

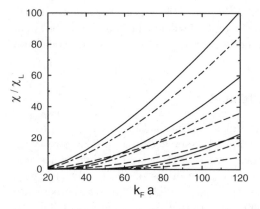

Fig. 6.4. Comparison between $k_\mathrm{F}a$ dependences of different contributions to the averaged susceptibility. Shown are the first-order interaction contribution $\chi_1^{\mathrm{ee,D}}$ for a local (δ-like) interaction (*full lines*), a screened interaction as in (6.31) (*dash–dotted lines*) and the contribution from the independent-particle model (*dashed lines*). Each set of curves belongs to the three temperatures $k_\mathrm{B}T/\Delta = 2, 4, 8$ (from the *top*)

The replacement (6.31) takes the short-range part of the interaction appropriately into account but does not account for possible contributions from the long-range tail of the interaction, which in the RPA behaves asymptotically as $\sim \sin(2k_\mathrm{F}r)/r^2$ [49]. Further semiclassical analysis shows that such contributions either are of lower order in $\hbar$ or do not survive the ensemble averaging, owing to the appearence of Friedel oscillations and long-range coupling between different trajectories.

6.4.3 Renormalization from Higher-Order Diagonal Contributions

As already seen for the diffusive case, higher-order diagrams are essential in the diagonal Cooper channel since they exhibit the same $\hbar$ dependence as the first-order term.

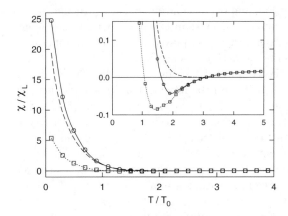

Fig. 6.5. Temperature dependence of the interaction contribution to the suscepti-bility (*solid line*) of an ensemble of squares at $k_F a = 50$ and $H = 0$. At low tempera-tures the susceptibility is dominated by the contribution of the nondiagonal channel from family $(1,1)$ and repetitions (*dashed line*), which exceeds that of the diagonal Cooper channel (*dotted line*); $k_B T_0 = \hbar v_F / 2\pi a$. The *inset* shows the minima of the *curves* on an enlarged scale (from [75]; ©1998 by the American Physical Society)

The main results of a numerical computation [75,313] (as described above) of the full interaction-induced average susceptibility (6.29) of the square billiard are summarized in Fig. 6.5. The zero-field diagonal contribution $\chi^{ee,D}(H{=}0)$ at $k_F a = 50$, including all orders in the local interaction, is given by the dotted curve. This curve exhibits three characteristic regimes [75]:

(i) At low temperature $\chi^{ee,D}$ is paramagnetic. It decays on a scale $k_B T_0 = \hbar v_F / 2\pi a$, which corresponds to a thermal cutoff length $L_{T_0} = 2a$. This behavior reflects the properties of the first-order term related to closed trajectories, as shown in Fig. 6.3a. Since their lengths are larger than $2a$, their contributions are exponentially suppressed for $L_T < 2a$.

(ii) In an intermediate temperature range, $\chi^{ee,D}$ is diamagnetic but small. This regime may be considered to represent the effects of the second-order terms: they are composed of closed paths consisting of two tra-jectories connected by interactions. Since there is no minimum length of these paths, the second-order term is less rapidly (and not exponentially) suppressed by temperature. The sign is opposite to the first-order term, explaining the the sign change in $\chi^{ee,D}$ near $T = T_0$.

(iii) In the limit of high temperatures $\chi^{ee,D}$ is again paramagnetic, although very small. In this regime, where $L_T \ll a$, third- and higher-order terms take over. They represent bulk-like contributions [315, 326] since orbits with more than two interactions may enclose flux without the assistance of the boundary.

However, the picture developed above needs refinement: in regime (i) the magnitude of $\chi^{\mathrm{ee,D}}$ is significantly smaller than the pure first-order diagonal contribution. Hence higher-order terms also play a role in this temperature range, pointing towards a renormalization mechanism.

The renormalization of the coupling constant λ_0 in diffusive systems is clearly expressed by means of (6.19). For ballistic dynamics such a clear-cut expression is lacking since one cannot employ properties such as (6.10). Nevertheless, the following picture may suggest a renormalization property of ballistic systems: at temperatures such that L_T is larger than the shortest periodic orbit the amplitudes entering into Σ^{D} in (6.28) are dominated by the $1/L_j^2$ dependence. This indicates that in an nth-order diagram, short path segments between two successive interactions are favored. This leads, on the one hand, to a tendency of the interaction events to cluster; on the other hand, paths contributing to the susceptibility have to enclose a significant amount of flux. This points towards the relevance of trajectories of the type shown in Fig. 6.3b. Nearby higher-order interaction events may be considered to act as one effective (renormalized) interaction. This rough picture can be put on a more rigorous footing by introducing a renormalization scheme, as was derived by Ullmo [75, 327]. This is achieved by regrouping the paths in the perturbation expansion (6.29) for $\Omega^{\mathrm{ee,D}}$ in such a way that short paths (between successive interaction events) which are not affected by the boundary are gathered into lower-order terms. For instance, the trajectory depicted in Fig. 6.3b then contributes to a first-order term in the reordered perturbation series where the short paths are incorporated in an effective interaction. On the basis of this renormalization scheme one obtains a resummed perturbation series for $\chi^{\mathrm{ee,D}}$ which can be interpreted as the same as the previous one, but now with the renormalized coupling constant $\lambda_0 = 1$ being replaced by the renormalized coupling constant [75]

$$\lambda(a) \simeq \frac{2}{\ln(k_F a)} \, . \tag{6.32}$$

In the semiclassical limit $\ln(k_F a) \gg 1$ the perturbation series in $\lambda(a)$ is now well converging and can be terminated after the first few terms. The reordered perturbation series for $\chi^{\mathrm{ee,D}}$ at $T < T_0$ is now dominated by its first-order contribution. In view of the smallness of $\lambda(a)$ it is now obvious that this first-order contribution is small compared to the original first-order term $\chi_1^{\mathrm{ee,D}}$ for $\lambda_0 = 1$. This explains the small amplitude of the $\chi^{\mathrm{ee,D}}$ in Fig. 6.5.

6.4.4 Nondiagonal Channel in Regular Systems

The use of the diagonal approximation in the semiclassical evaluation of $\hat{\Sigma}_\omega$ is in general justified for chaotic systems at not too low temperatures,[8] where

---

[8] In the limit where the temperature smoothing is on the order of the mean level spacing, corresponding to a cutoff at the Heisenberg time, off-diagonal terms

different orbits are assumed to be uncorrelated. However, nondiagonal terms in *regular* systems can persist after ensemble averaging and are of specific relevance to the magnetic response of interacting systems [75].

As discussed in detail in Sect. 2.1.3, periodic orbits in integrable systems are organized in families within which the action integral is constant. In general two (different) orbits of the same family cross in configuration space. If we associate the crossing point with the position of a first-order local interaction the dynamical phases of the nondiagonal pair of these orbits may cancel. This is illustrated in Fig. 6.3c for two orbits of the family (11) in the square.

This nondiagonal first-order contribution to the susceptibility from the family (1,1) and its higher repetitions k reads [75]

$$\frac{\langle \chi^{\text{ee,ND}} \rangle}{\chi_{\text{L}}} = -\frac{3k_{\text{F}}a}{2(\sqrt{2}\pi)^3} \sum_{k=1}^{\infty} \frac{1}{k} \frac{\text{d}^2 \mathcal{C}^2(k\varphi)}{\text{d}\varphi^2} R^2 \left(\frac{kL_{11}}{L_T} \right). \qquad (6.33)$$

Here, $\mathcal{C}(\varphi)$ is as defined in (4.64). As in the noninteracting case (Sect. 4.5.1), $\chi^{\text{ee,ND}}$ is linear in $k_{\text{F}}a$ and has a temperature scale given by R^2.

This nondiagonal channel, specific to regular systems, gains its importance from the fact that it is *not* renormalized by higher-order terms. A more detailed analysis based on phase space arguments shows that higher-order nondiagonal contributions are smaller by at least a factor $1/k_{\text{F}}a$.

For the case of square quantum dots the nondiagonal contribution of the family (1,1) and its repetitions is shown as the dashed curve in Fig. 6.5. It significantly exceeds the renormalized diagonal part and therefore gives the main contribution to the overall susceptibility χ^{ee}.

The existence of a nonrenormalized contribution in the nondiagonal channel for systems with integrable noninteracting dynamics leads to a qualitative difference from interacting systems with chaotic noninteracting behavior. This difference can be considered as a correlation effect since it stems from the effect of higher-order Cooper diagrams.

6.5 Comparison Between Integrable and Chaotic Structures

After the treatment of the magnetic response of clean quantum dots in Sect. 4, the study of the influence of disorder in Chap. 5, and the previous discussion of interaction effects, we are now in a position to make rather general final remarks about the orbital magnetism of mesoscopic quantum systems. In particular, we contrast systems with chaotic and regular dynamics in the corresponding noninteracting, clean counterparts.

---

may play a role in chaotic systems, similar to the effect of such terms in density correlators of noninteracting systems (see Sect. 2.2).

As discussed already in Sect. 4.4.2 on the level of a noninteracting model, the magnetic response of clean billiard-like microstructures differs for systems exhibiting families of periodic orbits (regular systems) and systems with only isolated periodic orbits (chaotic systems). The different $\hbar$ dependence of the density of states is reflected in a parametrically different dependence of the susceptibility on the Fermi energy or $k_F a$: the energy- or size-averaged susceptibility $\overline{\chi}$ of regular quantum dots scales $\sim k_F a$, while it is constant for chaotic systems[9] (see Table 6.1).

As discussed in the previous section, the quantum interaction contribution to $\overline{\chi}$ can be related (by using many-body perturbation theory) to the nature of the classical motion of the corresponding noninteracting system. In this case, families of periodic orbits give rise to nondiagonal terms in the Cooper channel. The related susceptibility contribution $\chi^{ee,ND}$ scales like $k_F a$ for regular systems. In chaotic systems, this channel does not contribute and the average susceptibility from the renormalized diagonal channel reads $\chi^{ee,D} \sim k_F a / \ln(k_F a)$. Hence, interacting and noninteracting terms in regular systems have the same $k_F a$ dependence, while in chaotic devices interaction effects yield the dominant contribution.

Table 6.1. Dependence on $k_F a$ of the noninteracting and interacting contributions to the magnetic response for billiard-like clean microstructures in the absence (chaotic case) and presence (regular case) of families of periodic orbits. $\overline{\chi}$ denotes the averaged (over energy) susceptibility.

	$\overline{\chi}/\chi_L$		χ/χ_L	
	Regular	Chaotic	Regular	Chaotic
Noninteracting	$(k_F a)$	$(k_F a)^0$	$(k_F a)^{3/2}$	$(k_F a)$
Interacting	$(k_F a)$	$(k_F a)/\ln(k_F a)$	$(k_F a)$	$(k_F a)/\ln(k_F a)$

To conclude, an ensemble of regular structures shows a magnetic response logarithmically larger than generic chaotic systems.

The parametric behavior for *individual* noninteracting systems was computed in Sects. 4.3.3 and 4.5.2. It is also summarized in Table 6.1. The leading-order interaction contribution for a single quantum dot is expected to be the same as for an ensemble because it is related to the trace of the diagonal part of Σ, which is a classical quantity. The trace integrals over oscillating contributions, which do exist in individual systems, merely give rise to $\hbar$ cor-

---

[9] Here, the discussion of the $k_F a$ dependence is based merely on the $\hbar$ dependence of individual orbit contributions. At rather low temperature, the common effect of the exponentially increasing number of longer periodic orbits in chaotic systems may change the overall k_F characteristics of the magnetic response, as was discussed in Sect. 4.3.3. The same holds true when including higher repetitions of orbits in regular systems.

rections to the contribution from Σ. A more detailed semiclassical study of such terms is, nevertheless, desirable.

With regard to temperature, both interacting and noninteracting semiclassical models yield an exponential decrease of the magnetic susceptibility if the corresponding thermal cutoff length L_T is of the order of the shortest periodic orbit.

In conclusion, the orbital magnetism represents an appropriate property for detecting differences in the quantum observables of classically integrable and chaotic ballistic systems. This holds true even in the presence of weak disorder, as was shown in Chap. 5.

6.6 Comparison with Experiment

The choice of square quantum dots as examples for studying orbital magnetism as well as for the investigation of disorder and interaction effects was partly motivated by the experiment of Lévy et al. [57]. To my knowledge, this experiment represents to date the sole measurement of the orbital magnetic response of an ensemble of ballistic quantum dots.

In this experiment the magnetic response of an array of 10^5 isolated mesoscopic billiards of approximately square geometry was measured. The squares were lithographically defined on a GaAs heterojunction. Their sizes are on average $a = 4.5\mu m$, but exhibit a wide variation of about 10 to 30% between the center and the border of the array. The two electron densities considered in the experiment were 10^{11} and 3×10^{11} cm^{-2}, corresponding to approximately 10^4 occupied levels per square. Therefore a semiclassical treatment is well justified.

The phase coherence length and elastic mean free path are estimated to be between 15 and 40 μm and between 5 and 10 μm, respectively. Hence, each square can be considered as phase-coherent and ballistic.

As the main experimental result in [57], a large paramagnetic peak at zero field was observed, two orders of magnitude larger than the Landau susceptibility, decreasing on a scale of approximately one flux quantum through each square. The results from the related semiclassical calculations (see (4.74) and (4.76)) and the full quantum calculations in the noninteracting model [45] are shown in Fig. 4.13b as the thick full and dashed lines, respectively (denoted by $\langle \chi \rangle$ in the figure). Corresponding curves from interaction contributions, not shown here, exhibit the same flux dependence. The theoretical curves decrease roughly on the same scale as observed in the experiment, see Fig. 1.5. (The offset in the semiclassical curve with respect to the quantum mechanical curve is due to the Landau susceptibility χ_L and additional effects from bouncing-ball orbits (see Sect. 4.5.3) not included in the semiclassical curve.)

More quantitatively, the measured paramagnetic susceptibility at $H = 0$ gave a value of approximately $100\chi_L$ (with an uncertainty of a factor of 4). For a temperature of 40 mK the factor $4\sqrt{2}/(5\pi)k_F a R_T^2(L_{11})$ from (4.75) for the clean noninteracting model gives zero-field susceptibility values of 60 and 170. In order to account for the effect of disorder we assume, as an unfavorable estimate, an elastic mean free path $l \simeq a$. The disorder-potential correlation length (5.4) can be estimated [72] to be of the order of $\xi/a \simeq 0.1$. We then obtain, with respect to the clean noninteracting case, a disorder reduction for the averaged susceptibility of $\overline{\langle \chi^s \rangle}/\overline{\chi} \simeq 0.4$ [72]. For $l \simeq a$ and the temperature used in the experiment, the disorder-induced contribution $\langle \chi^d \rangle$ discussed in Sect. 5.5.3 and proposed in [244] is small. This shows that the main features of the clean integrable system persist upon inclusion of the disorder parameters of the experiment.

The interaction contribution is roughly given by (6.33) from nondiagonal terms, which are a factor of 2 smaller than the result for the clean noninteracting case. Accounting for the disorder reduction and combining both contributions gives a susceptibility which is in line with the experimental values at low temperatures.

However, the experimentally observed decrease of the susceptibility with increasing temperature is less drastic than the theoretical predictions. In particular, the experimental T dependence is not exponential. The experiment shows a nonvanishing magnetic response at rather "high" temperatures of about 0.4 K, where theoretical approaches based on both noninteracting and interacting models yield a negligible susceptibility. Hence, the question of the experimental temperature behavior remains unsolved.

For completeness we note that the measurement of the persistent current of a single ballistic ring by Mailly, Chapelier, and Benoit [220] is in line with theory.

7. Concluding Remarks

In this book we have reviewed modern semiclassical approaches to mesoscopic quantum systems. We started by presenting basic semiclassical relations for spectral, thermodynamic, and transport properties. These expressions were further developed and applied to a number of different mesoscopic phenomena. We put particular emphasis on quantum transport and orbital magnetism. We treated in detail examples of current experimental relevance such as the conductivity of microcavities and antidot superlattices, the magnetic response of square quantum dots, and the persistent current of rings. The work on quantum transport was based on the model of noninteracting quasiparticles, while we included interaction effects in the study of the orbital magnetism.

The general picture which emerges from the semiclassical approach can be summarized as follows: in the semiclassical limit, spectral and transport quantities of an *individual* mesoscopic system are naturally decomposed into a smooth part and quantum corrections, which are usually of higher order in $\hbar$. Both the smooth and the oscillatory contributions can be consistently expanded in $\hbar$. However, for most of the mesoscopic phenomena treated here the corresponding leading-order term is sufficient. (The Landau diamagnetism represents an interesting exception since the classical magnetic response is zero.) The leading component of the smooth part yields the classical contribution, which may already contain nontrivial, system-specific information. Promiment examples are the resistivity anomaly of antidot arrays and photoabsorption cross sections. The oscillatory part is semiclassically represented as Fourier-type sums over classical paths, which are sensitive to disorder and temperature. As important examples, we discussed quantum oscillations of the magnetoconductivity observed in antidot experiments, which could be semiclassically attributed in a physically transparent way to interference effects of periodic orbits in the antidot lattice. Related quantum oscillations (as a function of the Fermi energy) were semiclassically predicted for the absorption of radiation in small ballistic particles and are precisely recovered in numerical quantum calculations [114]. The same holds true for the oscillations of the magnetic susceptibility in finite systems, extensively discussed for the example of the square quantum well.

With respect to *average* quantities the present situation of semiclassical theory differs for thermodynamic and transport properties. The computation of the magnetic response of an ensemble of quantum dots requires a canonical treatment. Comparison with numerical quantum calculations reveals that the average response is accurately described in terms of a semiclassical approximation of the variance of the particle number. This quantum "correction" due to the confinement dominates the small Landau diamagnetism in the mesoscopic regime.

Concerning averaged quantum corrections in transport, the situation is less clear. With regard to weak localization in ballistic cavities, we have shown how a semiclassical diagonal approximation yields a coherent-backscattering peak which can *qualitatively* explain quantum mechanical or experimental results. However, a semiclassical theory which can *quantitatively* describe average quantum transport is still missing.

One reason for the apparent differences between the semiclassical approaches to thermodynamic and transport properties is related to the way temperature enters. Averaged quantities can usually be expressed in terms of products of single-particle Green functions. Hence, a semiclassical evaluation of the (energy) average is in principle faced with the problem of a proper treatment of nondiagonal pairs of paths. For thermodynamic quantities such as orbital magnetism, each individual Green function is convoluted with the derivative of the Fermi function, yielding a cutoff in the lengths of contributing paths. Thus, in the mesoscopic regime ($k_B T > \Delta$) the semiclassical trace formulas for the magnetic response are well behaved and averaged quantum corrections are quantitatively obtained with high precision. In contrast to the thermodynamic quantities, temperature enters into semiclassical expressions for the averaged conductance as a suppression of pairs of paths, when the *difference* in the path lengths exceeds the temperature cutoff length. Hence, there is no clear-cut condition for a cutoff of long paths. This implies that orbits with periods of the order of or even larger than the Heisenberg time may enter and a proper inclusion of nondiagonal terms is required. However, as discussed at the end of Chap. 3, the inclusion of off-diagonal parts seems not to be sufficient to adequately account for quantum corrections. A refined theory would have to go beyond the semiclassical approximations used and would possibly have to account for nonclassical paths. Similar difficulties were reported for the weak-localization correction within the framework of the Kubo conductivity. Hence weak localization remains as a paradigm for the success and the open problems of semiclassics. In spite of the achievements of the present semiclassical formalism a quantitative and complete semiclassical transport theory is still lacking and would be highly desirable to complete our understanding of mesoscopic quantum transport.

A further general issue in quantum chaos which has guided our presentation refers to the question as to what extent the quantum properties of classically chaotic and nonchaotic systems differ. As shown, the semiclassical framework offers a suitable way to account for the different types of dynamics and is in this respect superior to random-matrix theory. The short-time dynamics usually exhibits nonuniversal features which may mask universal characteristics of chaotic systems. The role of the classical dynamics was illustrated in detail in the context of orbital magnetism. The presence of families of periodic orbits in regular systems makes their susceptibility parametrically larger than that of chaotic systems: within the model of independent particles the average susceptibility for a regular structure increases linearly in the large parameter $k_F a$, while it is independent of $k_F a$ for a chaotic system. This behavior has been quantitatively studied for various (integrable and chaotic) systems. As for the density of states, a general semiclassical theory for the transport and orbital magnetism of systems with mixed phase space dynamics is lacking.

Moreover, for quantum systems with purely chaotic classical dynamics the semiclassical theory of spectral correlations is far from being complete. Though commonly believed, the Bohigas–Giannoni–Schmit conjecture of random matrix theory for chaotic sytems remains to be proved. In this connection the deviations from random-matrix theory at finite energy are a topic of current interest which constitutes one interface between semiclassical approaches, supersymmetric techniques and random-matrix theory.

A large number of the phenomena in ballistic mesoscopic systems presented here can indeed, at least qualitatively, be described by using clean quantum billiards with independent particles as physical models. However, ballistic nanostructures are much richer than pure billiards: they represent complex condensed-matter systems including additional effects from applied magnetic fields, temperature, electron–phonon scattering, disorder, and electron–electron interactions.

At the very low temperature of most of the experiments considered here, electron–phonon scattering is strongly reduced and the phase-breaking length is much larger than the system size. In the example of transport through a finite antidot array (Sect. 3.2.4) the finite ℓ_ϕ could be incorporated in a rough way by means of a temperature-dependent cutoff length for the classical orbits involved. However, a rigorous theory which appropriately describes dissipation in the mesoscopic context would be highly desirable. This would also include questions of how finite-size effects in quantum coherent systems affect or may help to control dissipative processes. In particular, mesoscopic devices which are well controllable may serve as appropriate objects for studying the interplay between quantum chaos and dissipation [329].

Throughout this book we generally used perturbative concepts to include, for example, disorder and interaction effects or the effect of a small magnetic field. We note that the combination of classical perturbation theory and semiclassics is powerful because it is valid far beyond the range of quantum perturbation theory. This follows from the fact that semiclassical perturbation theory allows one to separate the effect of the perturbation on the phases from that on the classical dynamics. The latter is affected only on rather large classical magnetic-field scales.

Disorder effects are inherent in mesoscopic systems, though they can be suppressed in the ballistic regime. We reviewed in detail a semiclassical approach to accounting for weak-disorder effects in finite systems, considering observables related to the one-particle Green function (such as the magnetic susceptibility of a given structure) and to the two-particle Green function (such as the average susceptibility of an ensemble). We focused on the effect of smooth disorder in integrable structures. We showed that the interplay between finite size and a finite disorder correlation length results in a power-law damping and not an exponential damping of the average susceptibility. Indeed, the numerical and semiclassical studies of the effect of small residual disorder presented above show that, except for a possible reduction of the magnetic response, the description of the orbital magnetism as a geometrical effect in ballistic systems in terms of the orbits of the clean system remains essentially unaltered. To approach the diffusive limit, one has to take into account paths including scattering events. This can be rigorously performed within a semiclassical approach in the framework of a diagrammatic treatment of white-noise disorder: in this approach confinement effects in the crossover regime are included in the semiclassical Green function in terms of boundary-reflected paths.

A proper inclusion of electron–electron interactions in nanodevices is probably one of the most challenging programs in mesoscopics. We reviewed a semiclassical approach which rests upon a high-density perturbative expansion of interaction contributions to the thermodynamic potential. In reducing the average response of the interacting quantum problem to an essentially classical operator, which includes the classical probability for particles to return, the semiclassical method allows for a unified treatment of both diffusive and ballistic systems. Specifically, the orbital magnetic response in the ballistic regime is greatly enhanced over the Landau susceptibility owing to the combined effect of finite size and interactions. A parametric difference in the susceptibilities of regular and chaotic structures remains even when including interactions: the interaction contribution to $\overline{\chi}$ is proportional to $k_F a$ for regular geometries, while it is $\sim k_F a / \ln(k_F a)$ in the chaotic case. With regard to interactions, the role of families of periodic orbits is rather subtle, since they give rise to a nonrenormalizable additional nondiagonal contribution which is not present in chaotic systems. This treatment shows furthermore that the character of the classical dynamics of the noninteracting problem influences the quantum properties of the system of interacting electrons.

This approach can be regarded as one contribution to the general question of how chaos in a single-particle description is reflected in the properties of the full, interacting many-body problem including electron correlations. This open question fits into the general issue of how quantum chaos emerges in many-particle systems. This requires the development of methods from the field of nonlinear dynamics in many-body systems and includes the problem of higher-dimensional (quantum) chaos. Such methods seem appealing for attacking problems of correlated electrons and strongly interacting systems. Mesoscopic systems, which enable the confinement of interacting particles and the manipulation of their density, appear destined to be tools to investigate this challenging question of quantum chaos.

A. Appendices

A.1 Trace Integrals over Semiclassical Green Functions

In this section various useful trace integrals over products of Green functions are calculated in the stationary-phase approximation. They are, for example, helpful in the semiclassical computation of dynamic response functions (Sect. 2.4) and conductivities (Chap. 3).

A.1.1 Auxiliary Integrals

Definitions

Our starting point is two-dimensional integrals of the form

$$I_A^{++}(\boldsymbol{r}_1, \boldsymbol{r}_2) = \int d\boldsymbol{r}\; G^+(\boldsymbol{r}_2, \boldsymbol{r}; E + i\Gamma)\, A(\boldsymbol{r})\, G^+(\boldsymbol{r}, \boldsymbol{r}_1; E - \hbar\omega + i\Gamma), \quad \text{(A.1)}$$

$$I_A^{+-}(\boldsymbol{r}_1, \boldsymbol{r}_2) = \int d\boldsymbol{r}\; G^+(\boldsymbol{r}, \boldsymbol{r}_2; E + i\Gamma)\, A(\boldsymbol{r})\, G^-(\boldsymbol{r}_1, \boldsymbol{r}; E - \hbar\omega + i\Gamma). \quad \text{(A.2)}$$

The self-energy $\Gamma = \hbar/2\tau$ (assumed to be constant) accounts for disorder damping of the single-particle Green functions on the timescale of the relaxation time τ (see Chap. 5). For a semiclassical computation of the above integrals we express the Green functions according to (2.3) as sums over contributions G_t from classical paths $\mathcal{C}_t$ between $\boldsymbol{r}$ and $\boldsymbol{r}'$. In two dimensions, the G_t are of the form

$$G_t^+(\boldsymbol{r}', \boldsymbol{r}; E - \hbar\omega + i\Gamma)$$

$$= \frac{1}{i\hbar} \frac{1}{\sqrt{2i\pi\hbar}} D_t(\boldsymbol{r}, \boldsymbol{r}') \exp\left[\frac{i}{\hbar} S_t(E - \hbar\omega + i\Gamma) - i\eta_t \frac{\pi}{2}\right]. \quad \text{(A.3)}$$

For small ω and Γ we expand the action:

$$\frac{i}{\hbar} S_t(E - \hbar\omega + i\Gamma) \simeq \frac{i}{\hbar} S_t(E) - i\omega\tau_t - \frac{\tau_t}{2\tau}. \quad \text{(A.4)}$$

In the following, the functions $A(\boldsymbol{r})$ in (A.1) and (A.2) are assumed to vary slowly compared to the phases of order $S_t/\hbar$ in the Green function contributions. In a leading-order semiclassical approximation the operators $A(\boldsymbol{r})$ can then be replaced by their Weyl symbols and act as purely classical quantities, for example they commute with the Green functions.

The evaluation of the integrals (A.1) and (A.2) will be performed using local coordinates x along and y perpendicular to the trajectories $\bar{C}_t$ which are the stationary-phase solutions of the $\boldsymbol{r}$ integrals. The expansion of the action $S_t(\boldsymbol{r},\boldsymbol{r}')$ near the stationary point $\bar{\boldsymbol{r}}$ then yields [277]

$$G_t^+(\boldsymbol{r}',\boldsymbol{r};E-\hbar\omega+i\varGamma) = \frac{1}{i\hbar}\frac{1}{\sqrt{2\pi i\hbar}}D_t \tag{A.5}$$

$$\times \exp\left[\frac{i}{\hbar}\left(\bar{S}_t(\bar{\boldsymbol{r}},\boldsymbol{r}';E) + \frac{\partial S_t}{\partial y}y + \frac{1}{2}\frac{\partial^2 S_t}{\partial y^2}y^2\right) - \left(i\omega + \frac{1}{2\tau}\right)\bar{\tau}_t - i\bar{\eta}_t\frac{\pi}{2}\right].$$

It is convenient to express the action derivatives in terms of the monodromy matrix elements of the stationary path:

$$\frac{\partial^2 S_t}{\partial y^2} = \frac{m_{11}^t}{m_{12}^t} \quad ; \quad \frac{\partial^2 S_t}{\partial y'^2} = \frac{m_{22}^t}{m_{12}^t} \quad ; \quad \frac{\partial^2 S_t}{\partial y \partial y'} = -\frac{1}{m_{12}^t}. \tag{A.6}$$

The prefactor is approximated by

$$D_t(\boldsymbol{r},\boldsymbol{r}') \simeq \bar{D}_t(x,x') = \left(\frac{1}{m_{12}^t|\dot{x}\dot{x}'|}\right)^{1/2}. \tag{A.7}$$

Integrals of the Type I_A^{++}

Inserting (A.5) into (A.1) yields a double sum over pairs of paths $C_u(\boldsymbol{r}_1,\boldsymbol{r})$ and $C_v(\boldsymbol{r},\boldsymbol{r}_2)$. The stationary-phase condition for the integral I_A^{++} is (for small ω)

$$\frac{\partial}{\partial y}\left[\bar{S}_2(\boldsymbol{r},\boldsymbol{r}_2) + \bar{S}_1(\boldsymbol{r}_1,\boldsymbol{r})\right] = p_{y,1} - p_{y,2} \equiv 0 \tag{A.8}$$

with $(u,v) \equiv (1,2)$. This condition reduces the double sum to a single sum over combined smooth classical paths C_t from $\boldsymbol{r}_1$ to $\boldsymbol{r}_2$. The stationary-phase y integration amounts to replacing the prefactors $D_1 D_2$ by the new prefactor

$$\left(\frac{1}{m_{12}^1 m_{12}^2|\dot{x}_1\dot{x}^2\dot{x}_2|}\right)^{1/2}\left(\frac{2\pi i\hbar}{m_{22}^1/m_{12}^1 + m_{22}^2/m_{12}^2}\right)^{1/2} = \left(\frac{1}{|\dot{x}_1\dot{x}^2\dot{x}_2|}\frac{2\pi i\hbar}{m_{12}}\right)^{1/2}, \tag{A.9}$$

where m_{12} now denotes the off-diagonal element of the monodromy matrix of the full paths from $\boldsymbol{r}_1$ to $\boldsymbol{r}_2$.

The x integrations along the trajectories C_t involve the smooth functions $A(\bar{\boldsymbol{r}})$ and the velocity prefactor $1/|\dot{x}|$ from (A.9). They are suitably transformed into the time integrals

$$\langle A\rangle_t^{++}(\boldsymbol{r}_1,\boldsymbol{r}_2;\omega) = \int_0^{\tau(\boldsymbol{r}_2)-\tau(\boldsymbol{r}_1)} A[\boldsymbol{r}(t')]e^{-i\omega t'}\,dt'. \tag{A.10}$$

Combining (A.1), (A.5), (A.9), and (A.10), we retain a modified Green function structure

$$I_A^{++}(\boldsymbol{r}_1, \boldsymbol{r}_2) = \frac{1}{i\hbar} \sum_t \langle A \rangle_t^{++}(\boldsymbol{r}_1, \boldsymbol{r}_2; \omega) \, G_t^+(\boldsymbol{r}_2, \boldsymbol{r}_1; E + i\Gamma) \,. \qquad (A.11)$$

Integrals of the Type I_A^{+-}

Employing $G^-(\boldsymbol{r}', \boldsymbol{r}; E) = G^{+*}(\boldsymbol{r}, \boldsymbol{r}'; E)$, the integral (A.2) to be calculated is

$$I_A^{+-}(\boldsymbol{r}_1, \boldsymbol{r}_2) = \int d\boldsymbol{r} \, A(\boldsymbol{r}) \, G^+(\boldsymbol{r}, \boldsymbol{r}_2; E + i\Gamma) \, G^{+*}(\boldsymbol{r}, \boldsymbol{r}_1; E - \hbar\omega + i\Gamma) \,. \quad (A.12)$$

We exclude here the diagonal case where the two paths are identical (and where their phases cancel), which leads to the classical Weyl contribution discussed in Sect. 2.4. For $\boldsymbol{r}_1 \neq \boldsymbol{r}_2$ we can solve the integral again within the stationary-phase approximation, reducing the double sum over pairs of paths to a single sum. The corresponding stationary-phase condition reads

$$\frac{\partial}{\partial y}[\bar{S}_2(\boldsymbol{r}_2, \boldsymbol{r}) - \bar{S}_1(\boldsymbol{r}_1, \boldsymbol{r})] = p_{y,2} - p_{y,1} \equiv 0 \,, \qquad (A.13)$$

resulting in two classical paths beginning at $\boldsymbol{r}_1$ and $\boldsymbol{r}_2$, respectively, and ending at $\boldsymbol{r}$ with equal momenta. Therefore the shorter path lies entirely on the longer one. The quadratic fluctuations of the combined action functional are in this case

$$\frac{1}{2}\left(\frac{\partial^2 S_2}{\partial y^2} - \frac{\partial^2 S_1}{\partial y^2}\right) y^2 = \frac{1}{2}\left(\frac{m_{22}^2}{m_{12}^2} - \frac{m_{22}^1}{m_{12}^1}\right) y^2 \,. \qquad (A.14)$$

After stationary-phase integration, the new prefactor then reads (using (A.7))

$$\begin{aligned}
\frac{\tilde{D}_t(\boldsymbol{r}_1, \boldsymbol{r}_2)}{|\dot{x}|} &= D_1 D_2 \left(\frac{2\pi i\hbar}{m_{22}^2/m_{12}^2 - m_{22}^1/m_{12}^1}\right)^{1/2} \\
&= \left(\frac{1}{m_{12}^1 m_{12}^2 |\dot{x}_1 \dot{x}^2 \dot{x}_2|}\right)^{1/2} \left(\frac{2\pi i\hbar}{m_{22}^2/m_{12}^2 - m_{22}^1/m_{12}^1}\right)^{1/2} \\
&= \left(\frac{1}{|\dot{x}_1 \dot{x}^2 \dot{x}_2|} \frac{2\pi i\hbar}{m_{12}^1 m_{22}^2 - m_{12}^2 m_{22}^1}\right)^{1/2} \,. \qquad (A.15)
\end{aligned}$$

The x integration gives, after transformation to the time domain,

$$\langle A \rangle_t^{+-}(\omega, \tau) = \int_0^\infty A[\boldsymbol{r}(t')] e^{i\omega t'} e^{-t'/\tau} \, dt' \,. \qquad (A.16)$$

Here $\boldsymbol{r}(t'=0)$ is equal to $\boldsymbol{r}_1$ if $\boldsymbol{r}_1$ is located on the path between $\boldsymbol{r}_2$ and $\boldsymbol{r}$ and vice versa. $\boldsymbol{r}(t')$ runs along the (common) shorter path. By combining (A.5), (A.12), and (A.15) one finds

$$I_A^{+-}(\boldsymbol{r}_1, \boldsymbol{r}_2) = -\frac{1}{2\pi i\hbar^3} \sum_t \tilde{D}_t(\boldsymbol{r}_1, \boldsymbol{r}_2) \, \langle A \rangle_t^{+-} \exp\left[\frac{i}{\hbar} S_t(E) - \frac{\tau_t}{2\tau}\right]. \quad (A.17)$$

The remaining phase is now given by the action of the trajectory connecting r_1 and r_2 since we have combined G^+ and G^-. Depending on whether the longer path belongs to $G^+(E+i\Gamma)$ or $G^-(E-\hbar\omega+i\Gamma)$ one has to distinguish two cases for S_t for finite ω:

$$S_t(E) = \begin{cases} S_2(r_2, r_1; E) & \text{for } r_1 \text{ between } r_2 \text{ and } r, \\ -S_1(r_1, r_2; E) + \hbar\omega\tau_1(r_1, r_2) & \text{for } r_2 \text{ between } r_1 \text{ and } r. \end{cases} \quad (A.18)$$

A.1.2 Integrals of Products of Two Retarded Green Functions

For the semiclassical calculation of the matrix element sums arising in linear response theory (see Sect. 2.4) we have to compute trace integrals involving operators A and products of two retarded Green functions, as well as products of a retarded and an advanced Green function (in the noninteracting case, where the two-particle Green functions factorize). We begin with the case G^+G^+:

$$A_\Gamma^{++}(E;\omega)$$
$$\equiv \mathrm{Tr}[\hat{A}\, G^+(E+i\Gamma)\, \hat{A}\, G^+(E-\hbar\omega+i\Gamma)] \quad (A.19)$$
$$= \int dr' \int dr\, A(r')\, G^+(r', r; E+i\Gamma)\, A(r)\, G^+(r, r'; E-\hbar\omega+i\Gamma)\,.$$

Using (A.1) and (A.11), A_Γ^{++} can be represented (to leading order in $\hbar$) as

$$A_\Gamma^{++}(E;\omega) = \int dr'\, A(r')\, I_A^{++}(r', r') \quad (A.20)$$
$$\simeq \frac{1}{i\hbar} \sum_t \int dr'\, A(r')\langle A\rangle_t^{++}(r_1, r_2; \omega)\, G_t^+(r', r'; E+i\Gamma)$$

with $\langle A\rangle_t^{++}$ as defined in (A.10).

$A_\Gamma^{++}(E;\omega)$ can be considered as a generalized form of the trace integral representation of the density of states. Therefore, as we shall show below, the semiclassical computation of the integral will lead to a trace formula in terms of unstable periodic orbits, similar to the Gutzwiller trace formula (2.27) for the density of states.

In the following we focus on the *chaotic case*[1] assuming a phase space with only isolated periodic orbits. The trace integral (A.20) can be performed again by the stationary-phase approximation, following essentially the lines of Gutzwiller's semiclassical derivation of the density of states. The starting point is a representation of a closed (in configuration space) path $C_t(r', r')$ in local coordinates (x, y) of a nearby periodic orbit which results from the stationary-phase condition. Combining the factor arising from the stationary-phase integration in the y direction with the prefactor D_t of the Green function contribution G_t (see (A.7)) gives $\sqrt{2\pi i\hbar/m_{12}^{po}}/|\dot{x}|$. The velocity denominator is again used to transform the remaining x integral along

---

[1] A similar treatment is possible for the integrable case.

the orbit into a time correlation function of the classical function $A(\boldsymbol{r})$ along the periodic orbit. One finally obtains

$$A_\Gamma^{++}(E;\omega) \tag{A.21}$$

$$\simeq -\frac{1}{\hbar^2} \sum_{\text{po}} \sum_{j=1}^{\infty} \mathcal{C}_{\text{po}}^{++} \frac{e^{-j\tau_{\text{po}}/2\tau}}{\left|\det(\boldsymbol{M}_{\text{po}}^j - 1)\right|^{1/2}} \exp\left[ij\left(\frac{S_{\text{po}}}{\hbar} - \eta_{\text{po}}\frac{\pi}{2}\right)\right].$$

Here, $S_{\text{po}}(E,H)$ is the classical action, τ_{po} the period, and η_{po} the Morse index of a primitive periodic orbit. $\boldsymbol{M}_{\text{po}}$ denotes the monodromy matrix of the orbit and j counts higher repetitions. The operators $\hat{A}$ appear semiclassically via a Fourier-like transform of the correlation function of their classical analogs (Weyl symbols) A along each primitive periodic orbit:

$$\mathcal{C}_{\text{po}}^{++} = \int_0^{j\tau_{\text{po}}} dt\, e^{-i\omega t} \frac{1}{\tau_{\text{po}}} \int_0^{\tau_{\text{po}}} dt'\, A(t+t')A(t'). \tag{A.22}$$

A.1.3 Integrals of Products
of a Retarded and an Advanced Green Function

Here, we evaluate trace integrals involving operators $\hat{A}$ and a product of a retarded and an advanced Green function of the form

$$A_\Gamma^{+-}(E;\omega)$$

$$\equiv \text{Tr}[\hat{A}\, G^+(E + i\Gamma)\, \hat{A}\, G^-(E - \hbar\omega + i\Gamma)] \tag{A.23}$$

$$= \int d\boldsymbol{r}' \int d\boldsymbol{r}\, A(\boldsymbol{r})\, G^+(\boldsymbol{r},\boldsymbol{r}';E+i\Gamma)\, A(\boldsymbol{r}')\, G^-(\boldsymbol{r}',\boldsymbol{r};E-\hbar\omega+i\Gamma).$$

We express A_Γ^{+-}, analogously to the case of G^+G^+, by means of the auxiliary integral (A.2):

$$A_\Gamma^{+-}(E;\omega) = \int d\boldsymbol{r}'\, A(\boldsymbol{r}')\, I_A^{+-}(\boldsymbol{r}',\boldsymbol{r}') \tag{A.24}$$

with I_A^{+-} as defined in (A.17). Pairs of trajectories contributing to $A_\Gamma^{+-}(E;\omega)$ consist of paths of different length both beginning at $\boldsymbol{r}'$ with the longer one running through the point $\boldsymbol{r}'$ at least once more than the shorter one. Paths with the same length contribute to the Weyl part not considered here. Depending on whether the longer path belongs to $G^+(E + i\Gamma)$ or $G^-(E - \hbar\omega + i\Gamma)$, one has to distinguish for finite ω two contributions to the trace integral with the actions given in (A.18):

$$A_\Gamma^{+-}(E;\omega) = \frac{-1}{2\pi i\hbar^3} \sum_t \int d\boldsymbol{r}' A(\boldsymbol{r}')\, \tilde{D}_t(\boldsymbol{r}',\boldsymbol{r}')\langle A\rangle_t^{+-} e^{-\tau_t/2\tau} \tag{A.25}$$

$$\times \left\{ \exp\left[i\left(\frac{S_t(E)}{\hbar} - \eta_t\frac{\pi}{2}\right)\right] + \exp\left[-i\left(\frac{S_t(E)}{\hbar} - \eta_t\frac{\pi}{2} - \omega\tau_t\right)\right] \right\}.$$

Here S_t is now the action along the closed orbit segment $C_t(r', r')$. Assuming again chaotic phase space dynamics, the stationary-phase solutions of the above integral are dominated by contributions from closed orbits in the vicinity of a periodic orbit. Their actions are conveniently expanded in terms of the action S_{po} of the periodic orbit using its local coordinates:

$$
\begin{aligned}
S_t &= S_{\mathrm{po}} + \frac{1}{2} \left(\frac{\partial^2 S_t}{\partial y'^2} + 2\frac{\partial^2 S_t}{\partial y''\partial y'} + \frac{\partial^2 S_t}{\partial y''^2} \right)_{y'=y''=0} y'^2 \\
&\equiv \frac{1}{2} W(x') \, y'^2
\end{aligned}
\tag{A.26}
$$

with

$$
W(x) = \left[\frac{m_{11}(x) + m_{22}(x) - 2}{m_{12}(x)} \right]_{y'=y''=0} .
\tag{A.27}
$$

By performing the stationary-phase integral one obtains, instead of $\tilde{D}_t$ (see (A.15)), the new prefactor

$$
\bar{D}_t \equiv \frac{\tilde{D}_t}{\dot{x}'} \left(\frac{2\pi i\hbar}{W} \right)^{1/2} .
\tag{A.28}
$$

By denoting the monodromy matrix of the closed orbit by M and employing the relation $M_2 = M_1 \cdot M$ between the monodromy matrices M_2, M_1 of the (longer) path C_2 and the (shorter) path C_1, respectively, one can considerably simplify the product of $\tilde{D}_t$ (A.15) and $W^{-1/2}$ (A.27). After a few manipulations one finds

$$
\bar{D}_t = \frac{2\pi\hbar}{\dot{x}' \, |\det(M_{\mathrm{po}} - 1)|^{1/2}} .
\tag{A.29}
$$

The classical stability amplitude of the pair of open trajectories is thus just given by the corresponding periodic orbit. Hence, one obtains as the final result

$$
A_\Gamma^{+-}(E; \omega) \simeq \frac{1}{\hbar^2} \sum_{\mathrm{po}} C_{\mathrm{po}}^{+-}
\tag{A.30}
$$

$$
\times \sum_{j=1}^{\infty} \frac{\exp\left[(i\omega - 1/\tau)j\tau_{\mathrm{po}}/2\right] \cos\left[j(S_{\mathrm{po}}/\hbar - \omega\tau_{\mathrm{po}}/2 - \eta_{\mathrm{po}}\pi/2)\right]}{\left|\det(M_{\mathrm{po}}^j - 1)\right|^{1/2}} ,
$$

where the two contributions from (A.25) are included in a symmetrized way in the cos function. The C_{po}^{+-} are Fourier transforms of the correlation functions of the classical operator $A(r)$ along each primitive periodic orbit. They read

$$
C_{\mathrm{po}}^{+-} = \int_0^{\infty} dt \, e^{i\omega t - t/\tau} \frac{1}{\tau_{\mathrm{po}}} \int_0^{\tau_{\mathrm{po}}} dt' \, A(t + t')A(t') .
\tag{A.31}
$$

The trace integrals of the form $\mathrm{Tr}[A \, G^+ A \, G^-]$ calculated above yield the main contribution to the response functions discussed in Sect. 2.4.

A.2 Numerical Calculations for Susceptibilities of Square Quantum Dots

This appendix deals with the numerical procedure for calculating quantum mechanically the magnetic response of (disordered) square quantum wells at arbitrary magnetic field [45]. The quantum results are based on the diagonalization of the corresponding Hamiltonian and an efficient algorithm to calculate the canonical free energy. They are used to confirm the semiclassical results presented in Sect. 4.5 and Chap. 5.

Consider non-interacting spinless particles in a disordered square potential well $[-a/2, a/2]$ in a homogeneous magnetic field. Within the symmetric gauge $\boldsymbol{A} = H(-y/2, x/2, 0)$ the corresponding Hamiltonian in scaled units $\tilde{x} = x/a$ and $\tilde{E} = (ma^2/\hbar^2)E$ reads

$$\tilde{\mathcal{H}} = -\frac{1}{2}\left(\frac{\partial^2}{\partial \tilde{x}^2} + \frac{\partial^2}{\partial \tilde{y}^2}\right)$$
$$-i\pi\,\varphi\left(\tilde{y}\frac{\partial}{\partial \tilde{x}} - \tilde{x}\frac{\partial}{\partial \tilde{y}}\right) + \frac{\pi^2}{2}\varphi^2(\tilde{x}^2 + \tilde{y}^2) + V^{\mathrm{dis}}(\tilde{x}, \tilde{y})\,. \tag{A.32}$$

φ is the normalized flux defined as in (4.65) and V^{dis} a weak disorder potential.

A.2.1 Clean Case

In order to employ for $V^{\mathrm{dis}} \equiv 0$ the invariance of the Hamiltonian (A.32) with respect to rotations by $\pi, \pi/2$, we use linear combinations of plane waves which are eigenfunctions of the parity operators $\boldsymbol{P}_\pi, \boldsymbol{P}_{\pi/2}$, respectively. These combinations read, omitting the tilde,

$$\sqrt{2}[S_n(x)C_m(y) \pm iC_m(x)S_n(y)]\,, \qquad (P_\pi = -1)\,, \tag{A.33}$$

$$\left.\begin{array}{l}\sqrt{2}[C_n(x)C_m(y) \pm C_m(x)C_n(y)] \\ \sqrt{2}i\,[S_n(x)S_m(y) \pm S_m(x)S_n(y)]\end{array}\right\}\,, \qquad (P_\pi = +1)\,, \tag{A.34}$$

with $S_n(u) = \sin(n\pi u)$, n even, and $C_m(u) = \cos(m\pi u)$, m odd, obeying Dirichlet boundary conditions. In this representation the resulting matrix equation is real symmetric and decomposes into four blocks representing the different symmetry classes.

By diagonalization we calculated the first 3000 eigenenergies taking into account up to 2500 basis functions for each symmetry class. A typical energy level diagram of the symmetry class $(P_\pi, P_{\pi/2}) = (1, 1)$ as a function of the magnetic field is shown in Fig. 4.1.

A.2.2 Weak Disorder

In order to study disorder effects (Chap. 5) we use as a random potential

$$V^{\mathrm{dis}}(\boldsymbol{r}) = \sum_{j}^{N_i} \frac{u_j}{2\pi\xi^2} \exp\left(-\frac{(\boldsymbol{r}-\boldsymbol{R}_j)^2}{2\xi^2}\right), \qquad (A.35)$$

the sum of the Gaussian potentials (with impurity strength u_j) of N_i *independent* impurities located at points $\boldsymbol{R}_j$ with uniform probability. According to (A.44) and (A.45) the mean impurity strength and impurity density can be deduced from a given elastic mean free path. Because V^{dis} destroys the symmetry of the Hamiltonian of the clean case, the block structure is no longer preserved and the Hamiltonian becomes Hermitian.

A.2.3 Thermodynamics

One obtains the *grand canonical* susceptibility (see (4.2) and, for example, Fig. 4.2) from

$$\chi^{\mathrm{GC}}(\mu) = -\frac{g_{\mathrm{s}}}{a^2} \frac{\partial^2}{\partial H^2} \sum_{i=1}^{\infty} \frac{\epsilon_i}{1 + \exp[\beta(\epsilon_i - \mu)]}. \qquad (A.36)$$

Here g_{s} accounts for the spin degeneracy and ϵ_i denotes the single-particle energies.

In order to calculate ensemble-averaged quantities, such as the average susceptibility of an ensemble of square billiards, one often has to work in the *canonical* ensemble. At zero temperature the canonical free energy reduces to the total energy. Then the canonical susceptibility (4.17) is given as the sum over the curvatures of the N single-particle energies ϵ_i,

$$\chi(T=0) = -\frac{g_{\mathrm{s}}}{a^2} \sum_{i=1}^{N} \frac{\partial^2 \epsilon_i}{\partial H^2}. \qquad (A.37)$$

The susceptibility is therefore dominated by large paramagnetic singularities whenever the highest occupied state undergoes a level crossing with a state of a different symmetry class or a narrow avoided crossing with a state of the same symmetry. This makes $T = 0$ susceptibility spectra of quasi-integrable billiards (with nearly exact level crossings) or of separable systems with spectra composed of energy levels from different symmetry classes appear more erratic than those of chaotic systems, which possess stronger level repulsion [246].

The peaks for $T = 0$ are compensated once the next higher state at a (quasi) crossing is taken into account. Thus, these peaks disappear at finite temperature, when the occupation of nearly degenerate states becomes almost the same. Hence finite temperature regularizes the singular behavior of χ at $T=0$. The canonical susceptibility at finite temperature is given by

$$\chi = \frac{g_s}{a^2\beta} \frac{\partial^2}{\partial H^2} \ln Z_N(\beta) \ . \tag{A.38}$$

Here, the canonical partition function $Z_N(\beta)$ reads

$$Z_N(\beta) = \sum_{\{\alpha\}} \exp[-\beta E_\alpha(N)] \tag{A.39}$$

with

$$E_\alpha(N) = \sum_{i=1}^{\infty} \epsilon_i \, n_i^\alpha \quad , \quad N = \sum_{i=1}^{\infty} n_i^\alpha \ . \tag{A.40}$$

In the above equation the $n_i^\alpha \in \{0, 1\}$ label the occupation of the single-particle energy levels.

A brute-force numerical computation of the canonical partition function is usually extremely time-consuming at finite temperature. Thus, we first approximate the infinite sum in (A.39), which runs over all occupation distributions $\{\alpha\}$ for N electrons, by a finite sum $Z_N(M; \beta)$ over all possibilities for distributing N particles over the first M levels with $M \geq N$ sufficiently large. Following Brack et al. [330] we calculate $Z_N(M; \beta)$ recursively by employing the relation

$$Z_N(M; \beta) = Z_N(M - 1; \beta) + Z_{N-1}(M - 1; \beta) \exp(-\beta \epsilon_M) \tag{A.41}$$

and increase M until convergence of $Z_N(M, \beta)$ is obtained. The appropriate initial conditions are

$$Z_0(M; \beta) \equiv 1 \quad , \quad Z_N(N - 1; \beta) \equiv 0 \ . \tag{A.42}$$

This algorithm reduces the number of algebraic operations required to calculate Z_N significantly. Furthermore it is fast and accurate even for $k_B T/\Delta \simeq 10$. In such a temperature regime a direct calculation of Z_N is usually not possible.

A.3 Semiclassical and Quantum Results for Bulk Mean Free Paths

In this appendix we compare the semiclassical results of (5.13)–(5.15) for the elastic mean free path (MFP) in the ballistic regime with their counterparts obtained from quantum mechanical scattering theory and provide an expression for the transport MFP [72].

In the standard perturbative diagrammatic approach the effect of a weak random disorder potential is treated in the framework of the related Dyson equation for scattering using the self-consistent Born approximation. The resulting damping of the disorder-averaged single-particle Green function in a random potential is of the same exponential form as in (5.12) [281]. This is usually obtained by replacing the imaginary part of the self-energy in the

Green function after impurity averaging, which is associated with the inverse single-particle relaxation time, by the product of the density of states of the unperturbed system and the disorder parameters $n_i u^2$. Here n_i is the impurity density and u the mean potential strength of a scatterer. The resulting quantum mechanical elastic MFP l_{qm} which occurs in (5.12) is related to the total cross section σ by means of

$$\frac{1}{l_{qm}} = n_i \sigma , \qquad \sigma = \int d\Theta \, \sigma(\Theta) , \qquad (A.43)$$

with $\sigma(\Theta)$ being the partial cross section for scattering with an angle Θ.

For a Gaussian disorder potential of the form of (5.5) a calculation of the cross section can be performed analytically and the corresponding inverse MFP gives

$$\frac{1}{l_{qm}} = \frac{1}{l_\delta} I_0[2(k\xi)^2] \, e^{-2(k\xi)^2} . \qquad (A.44)$$

Here, I_0 is a modified Bessel function and

$$\frac{1}{l_\delta} = \frac{2\pi}{\hbar} \frac{n_i u^2}{v_F} N(0) = \frac{n_i u^2}{v_F} \frac{m}{\hbar^3} \qquad (A.45)$$

is the inverse MFP for the white-noise case of δ-like scatterers. v_F is the Fermi velocity and $N(0) = m/(2\pi\hbar^2)$ is the density of states of a 2DEG [281].

In order to compare l_{qm} with the semiclassical result, $l_{qm}(k\xi)$ can be expanded for large $k\xi$:

$$l_{qm}(k\xi) \simeq \sqrt{4\pi} \, (k\xi) \, l_\delta \left[1 - \frac{1}{16(k\xi)^2} \right] \qquad \text{for} \qquad k\xi \longrightarrow \infty . \ (A.46)$$

In the above equation the leading-order term is exactly the semiclassical MFP (5.13) for the Gaussian disorder model (5.5). The agreement between the semiclassical and diagrammatic approaches for the bulk can be related to the fact that the semiclassical treatment of disorder corresponds to the use of the eikonal approximation for each single scattering event, where the scattering potential is assumed to modify only the phases but not the trajectories themselves. The eikonal approximation is known to give the same results as the Born approximation for large $k\xi$.

For $\xi < \lambda_F$, the limit where the semiclassical description is no longer valid, the mean free path l_{qm} approaches l_δ. This means that (5.12) can still be used, but with the semiclassical l replaced by l_δ.

The transport MFP l_T is calculated quantum mechanically by including a factor $(1 - \cos\Theta)$ in the integral (A.43) for the scattering amplitude.[2] We find, for Gaussian disorder [72],

---

[2] This can be shown rigorously within diagrammatic perturbation theory including ladder diagrams [10].

$$\frac{1}{l_T} = \frac{1}{l_\delta} \left\{ I_0[2(k\xi)^2] - I_1[2(k\xi)^2] \right\} e^{-2(k\xi)^2} \tag{A.47}$$

$$\simeq \frac{1}{l_{qm}} \frac{1}{4(k\xi)^2} \qquad \text{for} \qquad k\xi \longrightarrow \infty. \tag{A.48}$$

According to this relation, l_T can be considerably larger than l_{qm} for $\lambda_F < \xi$. This shows that in the case of a confined system and smooth disorder, the system may behave ballistically even if the elastic MFP l is considerably smaller than the system size.

References

1. L.L. Sohn, Nature **394**, 131 (1998).
2. B.L. Altshuler, P.A. Lee, and R.A. Webb, *Mesoscopic Phenomena in Solids* (North-Holland, Amsterdam, 1991); B. Kramer and A. MacKinnon, Rep. Prog. Phys. **56**, 1469 (1993).
3. C. W. J. Beenakker and H. van Houten in *Semiconductor Heterostructures and Nanostructures*, Solid State Physics **44**, 1 (1991).
4. C. Rubio, N. Agraït, and S. Vieira, Phys. Rev. Lett. **76**, 2302 (1996).
5. R. Landauer, in *Coulomb and Interference Effects in Small Electronic Structures*, ed. by D.C. Glattli, M. Sanquer, and J. Trân Thanh Vân (Frontiers, Gif-sur-Yvette, 1994).
6. G. Bergmann, Phys. Rep. **107**, 1 (1984); P. A. Lee and T. V. Ramakrishnan, Rev. Mod. Phys. **57**, 287 (1985).
7. S. Chakravarty and A. Schmid, Phys. Rep. **140**, 193 (1986).
8. K. Efetov, *Supersymmetry in Disorder and Chaos* (Cambridge University Press, Cambridge, 1996).
9. Y. Imry, *Introduction to Mesoscopic Physics* (Oxford University Press, New York, 1997).
10. S. Datta, *Electronic Transport in Mesoscopic Systems* (Cambridge University Press, Cambridge, 1995).
11. D.K. Ferry and S.M. Goodnick, *Transport in Nanostructures* (Cambridge University Press, Cambridge, 1997).
12. *Mesoscopic Quantum Physics,* ed. by E. Akkermans, G. Montambaux, J.-L. Pichard, and J. Zinn-Justin (Elsevier, New York, 1995).
13. *Single Charge Tunneling*, ed. by H. Grabert and M.H. Devoret, NATO ASI Series B, **294** (Plenum, New York, 1992).
14. *Coulomb and Interference Effects in Small Electronic Structures*, ed. by D.C. Glattli, M. Sanquer, and J. Trân Thanh Vân (Frontiers, Gif-sur-Yvette, 1994).
15. *Quantum Dynamics of Submicron Structures*, ed. by H.A. Cerdeira, B. Kramer, and G. Schön, NATO ASI **291** (Kluwer, Dordrecht, 1994).
16. *Mesoscopic Electron Transport*, ed. by L.L. Sohn, L.P. Kouwenhoven, and G. Schön, NATO ASI Series E **345** (Kluwer, Dordrecht, 1997).
17. T. Dittrich, P. Hänggi, G.-L. Ingold, B. Kramer, G. Schön, and W. Zwerger, *Quantum Transport and Dissipation* (Wiley-VCH, Weinheim, 1997).
18. J. Math. Phys. **37** (10) (1996).
19. Chaos, Solitons & Fractals **8** (7,8) (1997).
20. L.P. Kouwenhoven, C.M. Marcus, P.L. McEuen, S. Tarucha, R.M. Westervelt, and N.S. Wingreen, Nato ASI Conference Proceedings, ed. by L.P. Kouwenhoven, G. Schön, and L.L. Sohn (Kluwer, Dordrecht, 1997).
21. K. Efetov, Adv. Phys. **32**, 53 (1983).

22. M.L. Mehta, *Random Matrices* (Academic Press, New York, 1991); F. Haake, *Quantum Signatures of Chaos*, (Springer, Berlin, Heidelberg, 1990).
23. O. Bohigas, in *Chaos and Quantum Physics*, ed. by M.-J. Giannoni, A. Voros, and J. Zinn-Justin (North-Holland, New York, 1991).
24. T. Guhr, A.M. Müller-Groeling, and H.A. Weidenmüller, Phys. Rep. **299**, 189 (1998).
25. C.W.J. Beenakker, Rev. Mod. Phys. **69**, 731 (1997).
26. M.C. Gutzwiller, *Chaos in Classical and Quantum Mechanics* (Springer, Berlin, Heidelberg, 1990).
27. A.M. Ozorio de Almeida, *Hamiltonian Systems: Chaos and Quantization* (Cambridge University Press, Cambridge, 1988).
28. *Chaos and Quantum Physics*, ed. by M.-J. Giannoni, A. Voros, and J. Zinn-Justin (North-Holland, New York, 1991).
29. L.E. Reichl, *The Transition to Chaos in Conservative and Classical Systems: Quantum Manifestations* (Springer, New York, 1992).
30. M. Brack and R.K. Bhaduri, *Semiclassical Physics*, Frontiers in Physics, **96**, (Addison-Wesley, Reading, 1997).
31. M.C. Gutzwiller, Am. J. Phys. **66**, 304 (1998).
32. H. Friedrich and D. Wintgen, Phys. Rep. **183**, 37 (1989).
33. Chaos **2** (1) (1992).
34. P. Cvitanović, *Classical and Quantum Chaos: A Cyclist Treatise*, "Das Buch", a webbook, accessible under http://www.nbi.dk/ChaosBook.
35. G. Ezra, K. Richter, G. Tanner, and D. Wintgen, J. Phys. B **24** L413 (1991); D. Wintgen, K. Richter, and G. Tanner, Chaos **2**, 19 (1992).
36. G. Tanner, K. Richter, and J.M. Rost, Rev. Mod. Phys. in press.
37. T. Chakraborty, *Quantum Dots: A survey of the properties of artificial atoms* (Elsevier, in press).
38. B. Shapiro, Physica A, **200**, 498 (1993).
39. Chaos **3** (4) (1993).
40. A.D. Stone, in *Mesoscopic Quantum Physics*, ed. by E. Akkermans, G. Montambaux, J.-L. Pichard, and J. Zinn-Justin (Elsevier, New York, 1995).
41. H.U. Baranger, in *Nano-Science and Technology*, ed. by G. Timp (Springer, Berlin, Heidelberg, 1997).
42. H.U. Baranger, R.A. Jalabert, and A.D. Stone, Phys. Rev. Lett. **65**, 2442 (1990).
43. H.U. Baranger, R.A. Jalabert, and A.D. Stone, Chaos **3**, 665 (1993).
44. A.M. Chang, H.U. Baranger, L.N. Pfeiffer, and K.W. West, Phys. Rev. Lett. **73**, 2111 (1994).
45. K. Richter, D. Ullmo, and R.A. Jalabert, Phys. Rep. **276**, 1 (1996).
46. O. Agam, B.L. Altshuler, and A.V. Andreev, Phys. Rev. Lett. **75**, 4389 (1995).
47. B.A. Muzykantskii and D.E. Khmelnitskii, JETP Lett. **62**, 76 (1995).
48. A.V. Andreev, O. Agam, B.D. Simons, and B.L. Altshuler, Phys. Rev. Lett. **76**, 3947 (1996).
49. T. Ando, A.B. Fowler, and F. Stern, Rev. Mod. Phys. **54**, 437 (1982).
50. D. Weiss and K. Richter, in *Mesoscopic Systems and Chaos: A Novel Approach*, ed. by G. Casati, H. Cerdeira, and S. Lundqvist (World Scientific, Singapore, 1994); D. Weiss and K. Richter, Phys. Blätter **51**, 171 (1995).
51. D. Weiss, K. Richter, A. Menschig, R. Bergmann, H. Schweizer, K. von Klitzing, and G. Weimann, Phys. Rev. Lett. **70**, 4118 (1993).
52. D. Weiss, G. Lütjering, and K. Richter, Chaos, Solitons & Fractals **8**, 1337 (1997).
53. T. Ericson, Phys. Rev. Lett. **5**, 430 (1960).
54. L.P. Kouwenhoven and C.M. Marcus, Physics World **11**, 35 (1998).

55. C.M. Marcus, A.J. Rimberg, R.M. Westervelt, P.F. Hopkings, and A.C. Gossard, Phys. Rev. Lett. **69**, 506 (1992).
56. H.U. Baranger, R.A. Jalabert, and A.D. Stone, Phys. Rev. Lett. **70**, 3876 (1993).
57. L.P. Lévy, D.H. Reich, L. Pfeiffer, and K. West, Physica B **189**, 204 (1993).
58. J.A. Melsen, P.W. Brouwer, K.M. Frahm. and C.W.J. Beenakker, Europhys. Lett. **35**, 7 (1996); Physica Scripta T **69**, 223 (1997).
59. I. Kostzin, D.L. Maslov, and P.M. Goldbart, Phys. Rev. Lett. **75**, 1735 (1995).
60. A. Altland and M.R. Zirnbauer, Phys. Rev. Lett. **76**, 3420 (1996).
61. A. Lodder and Yu.V. Nazarov, Phys. Rev. B **58**, 5783 (1998).
62. W. Ihra, M. Leadbeater, J.L. Vega, and K. Richter, Eur. Phys. J., submitted (1999).
63. M.F. Crommie, C.P. Lutz, and D.M. Eigler, Nature **363**, 524 (1993); E.J. Heller, M.F. Crommie, C.P. Lutz, and D.M. Eigler, Nature **369**, 464 (1994).
64. E.J. Heller, M.F. Crommie, and D.M. Eigler, Proceedings of the XIX ICPEAC, Whistler, BC, Canada (ed. by L.J. Dubé, J.B.A. Mitchell, J.W. McConkey, and C.E. Brion) AIP Conf. Proc. **360** (1995).
65. M.G.E. da Luz, A. Lupu Sax, and E.J. Heller, Phys. Rev. E **56**, 2496 (1997).
66. K. Richter, Europhys. Lett. **29**, 7 (1995).
67. G. Hackenbroich and F. von Oppen, Europhys. Lett. **29**, 151 (1995); Z. Phys. B **97**, 157 (1995).
68. G. Lütjering, K. Richter, D. Weiss, J. Mao, R.H. Blick, K. von Klitzing, and C.T. Foxon, Surf. Sci. **361/362**, 709 (1996).
69. R. A. Jalabert, K. Richter, D. Ullmo, Surf. Sci. **361/362**, 700 (1996).
70. K. Richter and B. Mehlig, Europhys. Lett. **41**, 587 (1998).
71. D. Ullmo, K. Richter, and R.A. Jalabert, Phys. Rev. Lett. **74**, 383 (1995).
72. K. Richter, D. Ullmo, and R.A. Jalabert, J. Math. Phys. **37**, 5087 (1996).
73. E. McCann and K. Richter, Europhys. Lett. **43**, 241 (1998).
74. D. Ullmo, K. Richter, H.U. Baranger, F. von Oppen, and R.A. Jalabert, Physica E **1**, 268 (1997).
75. D. Ullmo, H.U. Baranger, K. Richter, F. von Oppen, and R.A. Jalabert, Phys. Rev. Lett. **80**, 895 (1998).
76. N. Argaman, Y. Imry, and U. Smilansky, Phys. Rev. B **47**, 4440 (1993).
77. T. Dittrich, Phys. Rep. **271**, 267 (1996).
78. M.C. Gutzwiller, J. Math. Phys. **8**, 1979 (1967).
79. M.C. Gutzwiller, J. Math. Phys. **11**, 1791 (1970); J. Math. Phys. **12**, 343 (1971).
80. M.C. Gutzwiller, J. Math. Phys. **14**, 139 (1973).
81. B. Eckhardt and D. Wintgen, J. Phys. A **24**, 4335 (1991).
82. R. G. Littlejohn, J. Math. Phys. **31**, 2952 (1990).
83. S.C. Creagh, J.M. Robbins, and R.G. Littlejohn, Phys. Rev. A **42**, 1907 (1990).
84. O. Bohigas, M.-J. Giannoni, A.M. Ozorio de Almeida, and C. Schmit, Nonlinearity **8**, 203 (1995).
85. E.P. Wigner, Phys. Rev. **40**, 749 (1932).
86. B. Eckhardt, S. Fishman, K. Müller, and D. Wintgen, Phys. Rev. A **45**, 3531 (1992).
87. M. Seeley, Am. J. Math. **91**, 889 (1969).
88. M.V. Berry and M. Tabor, Proc. R. Soc. Lond. A **349**, 101 (1976).
89. M.V. Berry and M. Tabor, J. Phys. A **10**, 371 (1977).
90. R. Balian and C. Bloch, Ann. Phys. **60**, 401 (1970); Ann. Phys. **64**, 271 (1971); Ann. Phys. **69**, 76 (1972).
91. V.I. Arnold, *Mathematical Methods of Classical Mechanics*, Graduate Texts in Mathematics **60** (Springer, New York, 1989).

92. E.B. Bogomolny, Nonlinearity **5**, 805 (1992).
93. M.V. Berry and K.T. Mount, Rep. Prog. Phys. **35**, 315 (1972).
94. T. Szeredi, J.H. Lefebvre, and D.A. Goodings, Phys. Rev. Lett. **71**, 2891 (1993).
95. B. Georgeot and R.E. Prange, Phys. Rev. Lett. **74**, 2851 (1995); S. Fishman, B. Georgeot, and R.E. Prange, J. Phys. A **29**, 919 (1996).
96. U. Smilansky in *Mesoscopic Quantum Physics*, ed. by E. Akkermans, G. Montambaux, J.-L. Pichard, and J. Zinn-Justin (Elsevier, New York, 1995).
97. E.J. Heller in *Chaos and Quantum Physics*, ed. by M.-J. Giannoni, A. Voros, and J. Zinn-Justin (North-Holland, New York, 1991).
98. H. Schomerus and M. Sieber, J. Phys A **30**, 4537 (1997).
99. S. Tomsovic, N. Grinberg, and D. Ullmo, Phys. Rev. Lett. **75**, 4346 (1995).
100. G. Tanner, J. Phys. A **30**, 2863 (1997).
101. J. Main and G. Wunner, Phys. Rev. Lett. **82**, 3038 (1999).
102. A. Wirzba, habilitation thesis, University of Darmstadt (1997).
103. *Supersymmetry and Trance Formulae*, ed. by I.V. Lerner, J.P. Keating, and D.E. Khmelnitskii, NATO ASI Series B, **370** (Kluwer, New York, 1999).
104. J.P. Keating, in *Supersymmetry and Trance Formulae*, ed. by I.V. Lerner, J.P. Keating, and D.E. Khmelnitskii, NATO ASI Series B, **370** (Kluwer, New York, 1999).
105. B.L. Altshuler and B.I. Shklovskii, Zh. Eksp. Teor. Fiz. **91**, 220 (1986) [Sov. Phys. JETP **64**, 127 (1986)].
106. A.V. Andreev and B.L. Altshuler, Phys. Rev. Lett. **75**, 902 (1995).
107. M.V. Berry, Proc. R. Soc. Lond. A **400**, 229 (1985).
108. J.H. Hannay and A.M. Ozorio de Almeida J. Phys. **A17** 3429 (1984).
109. N. Argaman, F.-M. Dittes, E. Doron, J.P. Keating, Yu.A. Kitaev, M. Sieber, and U. Smilansky, Phys. Rev. Lett. **71**, 4326 (1993).
110. D. Cohen, H. Primack, and U. Smilansky, Ann. Phys. **264**, 108 (1998).
111. E.B. Bogomolny and J.P. Keating, Phys. Rev. Lett. **77**, 1472 (1996).
112. L.D. Landau and E.M. Lifshitz, *Statistical Physics* (Pergamon, Oxford, 1985).
113. L. Kouwenhoven, S. Jauhar, J. Orenstein, P.L. McEuwen, Y. Nagamune, J. Motoshia, and H. Sakaki, Phys. Rev. Lett. **73**, 3443 (1994).
114. B. Mehlig and K. Richter, Phys. Rev. Lett. **80**, 1936 (1998).
115. G. Czycholl and B. Kramer, Solid State Commun. **32**, 945 (1979).
116. Y. Imry and N.S. Shiren, Phys. Rev. B **33**, 7992 (1986).
117. N. Trivedi and D.A. Browne, Phys. Rev. B **38**, 9581 (1988).
118. D.J. Thouless and S. Kirkpatrick, J. Phys. C **14**, 235 (1981).
119. B. Reulet and H. Bouchiat, Phys. Rev. B **50**, 2259 (1994).
120. G.D. Mahan, *Many-Particle Physics* (Plenum, New York, 1990).
121. B. Mehlig and K. Richter, unpublished (1997).
122. M. Wilkinson, J. Phys. A **20**, 2415 (1987).
123. T. Prozen, Ann. Phys. **235**, 115 (1994).
124. B. Mehlig, D. Boosé, and K. Müller, Phys. Rev. Lett. **75**, 57 (1995).
125. D. Boosé, J. Main, B. Mehlig, and K. Müller, Europhys. Lett. **32**, 295 (1995).
126. B. Mehlig, Phys. Rev. B **55**, R10193 (1997).
127. B. Mehlig and K. Müller, unpublished (1996).
128. N. Argaman, Phys. Rev. Lett. **75**, 2750 (1995); Phys. Rev. B **53**, 7035 (1996).
129. A. Dellafiore, F. Matera, and D.M. Brink, Phys. Rev. A **51**, 914 (1995).
130. P. Gaspard and S. Jain, Pramana – J. Phys. **48**, 503 (1997); S. Jain and A.K. Pati, Phys. Rev. Lett. **80**, 650 (1998).
131. O. Zobey and G. Alber, J. Phys. B **26**, L539 (1993).
132. W.P. Halperin, Rev. Mod. Phys. **58**, 533 (1986).
133. V.V. Kresin, Phys. Rep. **220**, 1 (1992).

134. E.J. Austin and M. Wilkinson, J. Phys.: Cond. Mat. **5**, 8461 (1992); J. Phys.: Cond. Mat. **6**, 4153 (1994).
135. L.P. Gorkov and G.M. Eliashberg, Sov. Phys. JETP **21**, 940 (1965).
136. D. Heitmann and J.P. Kotthaus, Phys. Today **46**, 56 (1993); B. Meurer, D. Heitmann, and K. Ploog, Phys. Rev. Lett. **68**, 1371 (1992); K. Bollweg, T. Kurth, D. Heitmann, V. Gudmundsson, E. Vasiliadou, P. Grambow, and K. Eberl, Phys. Rev. Lett. **76** 2774 (1996).
137. J.B. Pieper and J.C. Price, Phys. Rev. Lett. **72**, 3586 (1994).
138. B. Reulet, M. Ramin, H. Bouchiat, and D. Mailly, Phys. Rev. Lett **75**, 124 (1995).
139. K. Richter, habilitation thesis, Universität Augsburg (1997).
140. K.B. Efetov, Phys. Rev. Lett. **76**, 1908 (1996).
141. C.M. Marcus, R.M. Westervelt, P.F. Hopkings, and A.C. Gossard, Chaos **3**, 643 (1993).
142. M.W. Keller, O. Millo, A. Mittal, D.E. Prober, and R.N. Sacks, Surf. Sci. **305**, 501 (1994).
143. M.J. Berry, J.H. Baskey, R.M. Westervelt, A.C. Gossard, Phys. Rev. B **50**, 8857 (1994).
144. R. Schuster, K. Ensslin, D. Wharam, S. Kühn, J. P. Kotthaus, G. Böhm, W. Klein, G. Tränkle, and G. Weimann, Phys. Rev. B **49**, 8510 (1994).
145. Y. Lee, G. Faini, and D. Mailly, Chaos, Solitons & Fractals **8**, 1325 (1997).
146. I.V. Zozoulenko, R. Schuster, K.-F. Berggren, and K. Ensslin, Phys. Rev. B **55**, R10209 (1997).
147. L. Christensson, H. Linke, P. Omling, P.E. Lindelof, I.V. Zozoulenko, and K.-F. Berggren, Phys. Rev. B **57**, 12306 (1998).
148. R.P. Taylor, R. Newbury, A.S. Sachrajda, Y. Feng, P.T. Coleridge, C. Dettmann, N. Zhu, H. Guo, A. Delage, P.J. Kelly, and Z. Wasilewski, Phys. Rev. Lett. **78**, 1952 (1997).
149. D.K. Ferry, R.A. Akis, D.P. Pivin, Jr., J.P. Bird, N. Holmberg, F. Badrieh, and D. Vasileska, Physica E **3**, 137 (1998).
150. A.S. Sachrajda, R. Ketzmerick, C. Gould, Y. Feng, P.J. Kelly, A. Delage, and Z. Wasilewski, Phys. Rev. Lett. **80**, 1948 (1998).
151. D. Weiss, P. Grambow, K. von Klitzing, A. Menschig, and G. Weimann, Appl. Phys. Lett. **58**, 2960 (1991).
152. D. Weiss, M.L. Roukes, A. Menschig, P. Grambow, K. von Klitzing, and G. Weimann, Phys. Rev. Lett. **66**, 2790 (1991).
153. R. Schuster, K. Ensslin, J.P. Kotthaus, M. Holland, and C. Stanley, Phys. Rev. B **47**, 6843 (1993).
154. R. Schuster, G. Ernst, K. Ensslin, M. Entin, M. Holland, G. Böhm, and W. Klein, Phys. Rev. B **50**, 8090 (1994).
155. J. Takahara, A. Nomura, K. Gamo, S. Takaoka, K. Murase, and H. Ahmed, Jpn. J. Appl. Phys. **34**, 4325 (1995).
156. O. Yevtushenko, G. Lütjering, D. Weiss, and K. Richter, Phys. Rev. Lett., submitted (1999).
157. T.M. Fromhold, L. Eaves, F.W. Sheard, M.L. Leadbeater, T.J. Foster, and P.C. Main, Phys. Rev. Lett. **72**, 2608 (1994); G. Müller, G. Boebinger, H. Mathur, L.N. Pfeiffer, and K.W. West, Phys. Rev. Lett. **75**, 2875 (1995).
158. T.M. Fromhold, P.B. Wilkinson, L. Eaves, F.W. Sheard, P.C. Main, M. Henini, M.J. Carter, N. Miura, and T. Takamasu, Chaos, Solitons & Fractals **8** 1381 (1997).
159. E.B. Bogomolny and D.C. Ruben, Europhys. Lett. **43**, 111 (1998).
160. E. Narimanov, A.D. Stone, and G.S. Boebinger, Phys. Rev. Lett. **78**, 4024 (1998).

161. J. Bardeen, Phys. Rev. Lett. **6**, 57 (1961).
162. E. Narimanov and A.D. Stone, Phys. Rev. B **57**, 9807 (1998).
163. D.S. Saraga and T.S. Monteiro, Phys. Rev. E **57**, 5252 (1998).
164. D. Weiss, Festkörperprobleme **31**, 341 (1991).
165. R. Schuster and K. Ensslin, Festkörperprobleme **34**, 195 (1994).
166. D. K. Ferry, Prog. Quant. Electr. **16**, 251 (1992); F. Nihey and K. Nakamura, Physica B **184**, 398 (1993).
167. P. Středa, J. Phys. C **15**, L717 (1982).
168. E. Vasiliadou, R. Fleischmann, D. Weiss, D. Heitmann, K. von Klitzing, T. Geisel, R. Bergmann, H. Schweizer, and C.T. Foxon, Phys. Rev. B **52**, R8658 (1995).
169. J. Blaschke, private communication (1998).
170. R. Fleischmann, T. Geisel, and R. Ketzmerick, Phys. Rev. Lett. **68**, 1367 (1992).
171. E.M. Baskin, G.M. Gusev, Z.D. Kvon, A.G. Pogosov, and M.V. Entin, JETP Lett. **55**, 678 (1992).
172. D. Weiss and K. Richter, Physica D**83**, 290 (1995).
173. D. Weiss, K. Richter, E. Vasiliadou, and G. Lütjering, Surf. Sci. **305**, 408 (1994).
174. G. Hackenbroich and F. von Oppen, Ann. Physik **5**, 696 (1996).
175. S. Ishizaka, F. Nihey, K. Nakamura, J. Sone, and T. Ando, Phys. Rev. B **51**, 9881 (1995); Jpn. J. Appl. Phys. **34**, 4317 (1995).
176. H. Silberbauer, J. Phys.: Cond. Mat. **4**, 7355 (1992); H. Silberbauer and U. Rössler, Phys. Rev. B **50**, 11911 (1994).
177. P. Rotter, M. Suhrke, and U. Rössler, Phys. Rev. B **54**, 4452 (1996).
178. R.B.S. Oakeshott and A. MacKinnon, J. Phys. C **5**, 6991 (1993); J. Phys. C **6**, 1519 (1994).
179. S. Uryu and T. Ando, Phys. Rev. B **53**, 13613 (1996).
180. T. Ando, S. Uryu, S. Ishizaka, and T. Nakanishi, Chaos, Solitons & Fractals **8**, 1057 (1997).
181. M. Suhrke and P. Rotter, in *Theory of Transport Properties of Semiconductor Nanostructures*, ed. by E. Schöll (Chapman and Hall, London, 1997).
182. P. Rotter, U. Rössler, M. Suhrke, R. Schuster, R. Neudert, K. Ensslin, and J.P. Kotthaus, in *The Physics of Semiconductors*, ed. by M. Scheffler and R. Zimmermann (World Scientific, Singapore, 1996).
183. H. Silberbauer, Ph.D. thesis, University of Regensburg (1994).
184. I.V. Zozoulenko and K.-F. Berggren, Phys. Rev. B **54**, 5823 (1996).
185. G.M. Gusev, P. Basmaji, D.I. Lubyshev, J.C. Portal, L.V. Litvin, Yu.V. Nastaushev, and A.I. Toropov, Solid-State Electronics **37**, 1231 (1994).
186. H.U. Baranger, D.P. DiVincenzo, R.A. Jalabert, and A.D. Stone, Phys. Rev. B **44**, 10637 (1991).
187. R. Landauer, Phil. Mag. **21**, 863 (1970).
188. R. Blümel and U. Smilansky, Phys. Rev. Lett. **60**, 477 (1988).
189. C.D. Schwieters, J.A. Alford, and J.B. Delos, Phys. Rev. B **54**, 10652 (1996).
190. W.A. Lin and R.V. Jensen, Phys. Rev. B **53**, 3638 (1996).
191. H. Ishio and J. Burgdörfer, Phys. Rev. B **51**, 2013 (1995).
192. M. Schreier, K. Richter, G.-L. Ingold, and R.A. Jalabert, Euro. Phys. J. B **3**, 387 (1998).
193. E. Doron, U. Smilansky, and A. Frenkel, Physica **D50** 367 (1991); Phys. Rev. Lett. **65**, 3072 (1990).
194. P. Gaspard and S.A. Rice, J. Chem. Phys. **90**, 2225, 2242, 2255 (1989).
195. D.K. Ferry and S.M. Goodnick, *Transport in Nanostructures*, Chap. 5.7 (Cambridge University Press, Cambridge, 1997).

196. I.V. Zozoulenko, F.A. Maaø, and E.H. Hauge, Phys. Rev. B **53**, 7975, 7987 (1996); Phys. Rev. B **54**, 4710 (1997).
197. R. Ketzmerick, Phys. Rev. B **54**, 10841 (1996).
198. H.U. Baranger and P. Mello, Phys. Rev. Lett. **73**, 142 (1994); R.A. Jalabert, J.-L. Pichard, and C.W.J. Beenakker, Europhys. Lett. **27**, 255 (1994).
199. Z. Pluhař, H.A. Weidenmüller, J.A. Zuk, and C.H. Lewenkopf, Phys. Rev. Lett. **73**, 2115 (1994).
200. K. Efetov, Phys. Rev. Lett. **74**, 2299 (1995).
201. P.A. Mello and H.U. Baranger, cond-mat/9812225 at http://xxx.lanl.gov/ (1998).
202. D. Ullmo and H. Baranger, private communication.
203. K. Richter, unpublished.
204. H.U. Baranger and A.D. Stone, Phys. Rev. B **40**, 8169 (1989).
205. I.L. Aleiner and A.I. Larkin, Phys. Rev. B **54**, 14423 (1996); Phys. Rev. E **55**, R1234 (1997); Chaos, Solitons & Fractals **8**, 1179 (1997).
206. B.V. Chirikov, F.M. Izrailev, and D.L. Shepelyanskii, Sov. Sci. Rev. C **2**, 209 (1981).
207. R.S. Whitney, I.V. Lerner, and R.A. Smith, cond-mat/9902328 at http://xxx.lanl.gov/ (1999).
208. S. Hikami, Phys. Rev. B **24**, 2671 (1981).
209. Y. Takane and K. Nakamura, J. Phys. Soc. Japan, **66**, 2977 (1997).
210. L.P. Lévy, G. Dolan, J. Dunsmuir, and H. Bouchiat, Phys. Rev. Lett. **64**, 2074 (1990).
211. V. Chandrasekhar, R.A. Webb, M.J. Brady, M.B. Ketchen, W.J. Gallagher, and A. Kleinsasser, Phys. Rev. Lett. **67**, 3578 (1991).
212. M. Büttiker, Y. Imry, and R. Landauer, Phys. Lett. A **96**, 365 (1983).
213. H.F. Cheung, Y. Gefen, and E.K. Riedel, IBM J. Res. Develop. **32**, 359 (1988).
214. H. Bouchiat and G. Montambaux, J. Phys. (Paris) **50**, 2695 (1989); G. Montambaux, H. Bouchiat, D. Sigeti, and R. Friesner, Phys. Rev. B **42**, 7647 (1990).
215. Y. Imry, in *Coherence Effects in Condensed Matter Systems*, ed. by B. Kramer (Plenum, New York, 1991).
216. A. Schmid, Phys. Rev. Lett. **66**, 80 (1991); F. von Oppen and E.K. Riedel, **66**, 84 (1991); B.L. Altshuler, Y. Gefen, and Y. Imry, **66**, 88 (1991).
217. P. Mohanty, E.M.Q. Jariwala, M.B. Ketchen, and R.A. Webb, in *Quantum Coherence and Decoherence*, ed. by K. Fujikawa and Y.A. Ono (Elsevier Science, Amsterdam, 1996).
218. A. Müller-Groeling and H.A. Weidenmüller, Phys. Rev. B**49**, 4752 (1994).
219. U. Eckern and P. Schwab, Adv. Phys. **44**, 387 (1995).
220. D. Mailly, C. Chapelier, and A. Benoit, Phys. Rev. Lett. **70**, 2020 (1993).
221. R.A. Jalabert, K. Richter, and D. Ullmo, in *Coulomb and Interference Effects in Small Electronic Structures*, ed. by D.C. Glattli, M. Sanquer, and J. Trân Thanh Vân (Frontiers, Gif-sur-Yvette, 1994).
222. O. Agam, J. Phys. I (France) **4**, 697 (1994).
223. F. von Oppen, Phys. Rev. B **50**, 17151 (1994).
224. S.E. Apsel, J.W. Emmert, J. Deng, and L.A. Bloomfield, Phys. Rev. Lett. **76**, 1441 (1996).
225. L.D. Landau, Z. Phys. **64**, 629 (1930).
226. J. H. van Leeuwen, J. Phys. (Paris) **2**, 361 (1921).
227. R.E. Peierls, *Quantum Theory of Solids* (Oxford University Press, Oxford, 1964); *Surprises in Theoretical Physics* (Princeton University Press, Princeton, NJ, 1979).
228. D. Shoenberg, Proc. R. Soc. **170 A**, 341 (1939).

229. D.A. van Leeuwen, Ph.D. thesis, University of Leiden (1993).
230. J.M. van Ruitenbeek and D.A. van Leeuwen, Mod. Phys. Lett. B **7**, 1053 (1993).
231. A. Papapetrou, Z. Phys. **107**, 387 (1937); L. Friedman, Phys. Rev. **134 A**, 336 (1964); S.S. Nedorezov, Zh. Eksp. Teor. Fiz. **64**, 624 (1973) [Sov. Phys. JETP **37**, 317 (1973)].
232. R.B. Dingle, Proc. R. Soc. **212 A**, 47 (1952).
233. R.V. Denton, Z. Phys. **265**, 119 (1973).
234. R. Németh, Z. Phys. B **81**, 89 (1990).
235. E.N. Bogacheck and G.A. Gogadze, Pis'ma Zh. Eksp. Teor. Fiz. **63**, 1839 (1972) [Sov. Phys. JETP **36**, 973 (1973)].
236. D.B. Bivin and J.W. McClure, Phys. Rev. B **16**, 762 (1977).
237. W. Lehle, Yu.N. Ovchinnikov, and A. Schmid, Ann. Physik **6**, 487 (1997).
238. B.K. Jennings and R.K. Bhaduri, Phys. Rev. B **14**, 1202 (1976).
239. M. Robnik, J. Phys. A **19**, 3619 (1986).
240. M. Antoine, thesis, Université Paris VI (1991).
241. J.M. van Ruitenbeek, Z. Phys. D **19**, 247 (1991); J.M. van Ruitenbeek and D.A. van Leeuwen, Phys. Rev. Lett. **67**, 641 (1991).
242. K. Rezakhanlou, H. Kunz, and A. Crisanti, Europhys. Lett. **16**, 629 (1991).
243. F. von Oppen and E.K. Riedel, Phys. Rev. B **48**, 9170 (1993).
244. Y. Gefen, D. Braun, and G. Montambaux, Phys. Rev. Lett. **73**, 154 (1994).
245. M.V. Berry and J.P. Keating, J. Phys. A **27**, 6167 (1994).
246. K. Nakamura and H. Thomas, Phys. Rev. Lett. **61**, 247 (1988).
247. H. Mathur, M. Gökçedag, and A.D. Stone, Phys. Rev. Lett. **74**, 1855 (1995).
248. J.P. Eisenstein, H.L. Störmer, V. Narayanamurti, A.Y. Cho, A.C. Gossard, and C.W. Tu, Phys. Rev. Lett. **55**, 875 (1985).
249. I. Meinel, T. Hengstmann, D. Grundler, D. Heitmann, W. Wegscheider, and M. Bichler, Phys. Rev. Lett. **82**, 819 (1999).
250. O. Bohigas, S. Tomsovic, and D. Ullmo, Phys. Rep. **223**, 43 (1993).
251. S.D. Prado, M.A.M. de Aguiar, J.P. Keating, and R. Egydio de Carvalho, J. Phys. A **27**, 6091 (1994).
252. R. Kubo, J. Phys. Soc. Japan **19**, 2127 (1964).
253. Y. Imry, in *Directions in Condensed Matter Physics*, ed. by G. Grinstein and G. Mazenko (World Scientific, Singapore, 1986).
254. R.E. Prange, Phys. Rev. Lett. **78**, 2280 (1997).
255. M.V. Berry and J.P. Keating, Proc. R. Soc. Lond. A **437**, 151 (1992).
256. R.A. Serota, Solid State Commun. **84**, 843 (1992).
257. A. Szafer and B.L. Altshuler, Phys. Rev. Lett. **70**, 587 (1993).
258. R. Jensen, Chaos **1**, 101 (1991).
259. W.H. Miller, Adv. Chem. Phys. **9**, 48 (1974).
260. K. Richter and B. Mehlig, unpublished.
261. T. Dittrich, B. Mehlig, H. Schanz, and U. Smilansky, Chaos, Solitons & Fractals **8**, 1205 (1997).
262. H.J. Stöckmann, private communication (1997).
263. A.M. Ozorio de Almeida, in *Quantum Chaos and Statistical Nuclear Physics*, Lecture Notes in Physics **263**, ed. by T. Seligman (Springer, Berlin, Heidelberg, 1986).
264. S. Creagh, Ann. Phys. (New York) **248**, 60 (1996).
265. I.S. Gradshteyn and I.M. Ryzhik, *Tables of Integrals, Series and Products* (Academic Press, New York, 1980).
266. V.N. Kondratyev and H.O. Lutz, Phys. Rev. Lett. **81**, 4508 (1998).
267. V. Fock, Z. Phys. **47**, 446 (1928).

268. I.O. Kulik, Zh. Eksp. Teor. Fiz. **58** 2171 (1970) [Sov. Phys. JETP **31**, 1172 (1970)]; ZhETF Pis. Red. **11** 407 (1970) [JETP Lett. **11**, 275 (1970)].
269. F. von Oppen, Ph.D. thesis (University of Washington, 1993).
270. J.B. Keller and S.I. Rubinow, Ann. Phys. **9**, 24 (1960).
271. U. Sivan and Y. Imry, Phys. Rev. Lett. **61**, 1001 (1988).
272. D. Yoshioka, J. Phys. Soc. Japan **62**, 3198 (1993).
273. J. Hajdu and B. Shapiro, Europhys. Lett. **28**, 61 (1994).
274. M. Robnik, in *Nonlinear Phenomena and Chaos*, ed. by S. Sarkar (Adam Hilger, Bristol, 1986).
275. J. Blaschke and M. Brack, Phys. Rev. A**56**, 182 (1997).
276. E. Gurevich and B. Shapiro, J. Phys. I (France) **7**, 807 (1997).
277. E.B. Bogomolny, Physica D **31**, 169 (1988).
278. A. Altland and Y. Gefen, Phys. Rev. Lett. **71**, 3339 (1993); Phys. Rev. B **51**, 10671 (1995).
279. U. Sivan and Y. Imry, Phys. Rev. B **35**, 6074 (1987).
280. S. Das Sarma and F. Stern, Phys. Rev. B **32**, 8442 (1988).
281. A.A. Abrikosov, L.P. Gorkov, and I.E. Dzyaloshinski, *Methods of Quantum Field Theory in Statistical Physics* (Prentice-Hall, Englewood Clifs, 1963).
282. E. McCann and K. Richter, Phys. Rev. B **59**, 13026 (1999).
283. K. Richter, D. Ullmo, and R.A. Jalabert, Phys. Rev. B **54**, R5219 (1996).
284. A.D. Mirlin, E. Altshuler, and P. Wölfle, Ann. Physik **5**, 281 (1996).
285. J.A. Nixon and J.H. Davies, Phys. Rev. B **41**, 7929 (1990).
286. A.D. Stone, in *Physics of Nanostructures*, ed. by J.H. Davies and A.R. Long (IOP Publishing, Bristol, 1992).
287. S. Doniach and E.H. Sondheimer, *Green's Functions for Solid State Physicists* (Addison-Wesley, Reading, 1978).
288. M.E. Raikh and T.V. Shahbazyan, Phys. Rev. B **47**, 1522 (1993).
289. K. Richter, unpublished; see also [284].
290. F. Wegner, Z. Phys. B **51**, 279 (1983).
291. A.M. Zagoskin, S.N. Rashkeev, R.I. Shekhter, and G. Wendin, cond-mat/9404077 at http://xxx.lanl.gov/ (1994).
292. S. Oh, A.Yu. Zyuzin, and A. Serota, Phys. Rev. B **44**, 8858 (1991).
293. A. Raveh and B. Shapiro, Europhys. Lett. **19**, 109 (1992); B.L. Altshuler, Y. Gefen, Y. Imry, and G. Montambaux, Phys. Rev. B **47**, 10340 (1993).
294. P.G. de Gennes and M. Tinkham, Physics **1**, 107 (1964).
295. A. Altland, Y. Gefen, and G. Montambaux, Phys. Rev. Lett. **76**, 1130 (1996).
296. O. Agam and S. Fishman, Phys. Rev. Lett. **76**, 726 (1996).
297. Ya.M. Blanter, A.D. Mirlin, and B.A. Muzykantskii, Phys. Rev. Lett. **80**, 4161 (1998); V. Tripathi and D.E. Khmelnitskii, Phys. Rev. B **58**, 1122 (1998).
298. M. Büttiker, A. Prêtre, and H. Thomas, Phys. Rev. Lett. **70**, 4114 (1993); D.Z. Liu, B.Y.-K. Hu, C.A. Stafford, and S. Das Sarma, Phys. Rev. B **50**, 5799 (1994); G. Cuniberti, M. Sassetti, and B. Kramer, Phys. Rev. B **57**, 1515 (1998); Ya.M. Blanter, F.W.J. Hekking, and M. Büttiker, Phys. Rev. Lett. **81**, 1925 (1998) and references therein.
299. J.H.F. Scott-Thomas, S.B. Field, M.A. Kastner, D.A. Antoniadis, and H.I. Smith, Phys. Rev. Lett. **62**, 583 (1989); M. Kastner, Rev. Mod. Phys. **64**, 849 (1992).
300. A.M. Chang, H.U. Baranger, L.N. Pfeiffer, K.W. West, and T.Y. Chang, Phys. Rev. Lett. **76**, 1695 (1996); J.A. Folk, S.R. Patel, S.F. Gogijn, A.G. Huibers, S.M. Cronenwett, C.M. Marcus, K. Campman, and A.C. Gossard, Phys. Rev. Lett. **76**, 1699 (1996).

301. U. Sivan, R. Berkovits, Y. Aloni, O. Prus, A. Auerbach, and G. Ben-Yosef, Phys. Rev. Lett. **77**, 1123 (1996); F. Simmel, T. Heinzel, and D.A. Wharam, Europhys. Lett. **38**, 123 (1997); S.R. Patel, S.M. Cronenwett, D.R. Stewart, A.G. Huibers, C.M. Marcus, C.I. Druöz, J.S. Harris, Jr., K. Campman, and A.C. Gossard, Phys. Rev. Lett. **80**, 4522 (1998); F. Simmel, D. Abusch-Magder, D.A. Wharam, M.A. Kastner, and J.P. Kotthaus, Phys. Rev. B **59**, R10441 (1999).

302. O. Agam, N.S. Wingreen, B.L. Altshuler, D.C. Ralph, and M. Tinkham, Phys. Rev. Lett. **78**, 1956 (1997).

303. J.M. Verbaarschot, H.A. Weidenmüller, and M.R. Zirnbauer, Phys. Rep. **129**, 367 (1985).

304. R.A. Jalabert, A.D. Stone, and Y. Alhassid, Phys. Rev. Lett. **68**, 3468 (1992).

305. R. Berkovits and B.L. Altshuler, Phys. Rev. B **55**, 5297 (1997); Y.M. Blanter, A.D. Mirlin, and B. Muzykantskii, Phys. Rev. Lett. **78**, 24449 (1997); R. Vallejos, C.H. Lewenkopf, and E.R. Mucciolo, Phys. Rev. Lett. **81**, 677 (1998); M. Stopa, Physica B **251**, 228 (1998); R. Berkovits, Phys. Rev. Lett. **81**, 2128 (1998); A. Cohen, K. Richter, and R. Berkovits, Phys. Rev. B **60**, 2536 (1999); S. Levit and D. Orgad, Phys. Rev. B **60**, 5549 (1999); P.N. Walker, G. Montambaux, and Y. Gefen, Phys. Rev. Lett. **82**, 5329 (1999).

306. K.H. Ahn, K. Richter, and I.H. Lee, Phys. Rev. Lett., in press (1999); K.H. Ahn and K. Richter, Physica E, in press (1999).

307. D. Shepelyansky, Phys. Rev. Lett. **73**, 2607 (1994).

308. Y. Imry, Europhys. Lett. **30**, 405 (1995); K. Frahm, A. Müller-Groeling, J.L. Pichard, and D. Weinmann, Europhys. Lett. **31**, 169 (1995); F. von Oppen, T. Wettig, and J. Müller, Phys. Rev. Lett. **76**, 491 (1996).

309. R. Römer and M. Schreiber, Phys. Rev. Lett. **78**, 515 (1997).

310. K. Frahm, A. Müller-Groeling, J.L. Pichard, and D. Weinmann, Phys. Rev. Lett. **78**, 4889 (1997).

311. Ann. Physik **7** (5–6) (1998); proceedings of the conference *Localization 99*, Ann. Physik, in press (1999).

312. I.L. Aleiner and L.I. Glazman, cond-mat/9710195 at http://xxx.lanl.gov (1997); Y. Takane, J. Phys. Soc. Japan **67**, 3003 (1998).

313. D. Ullmo, H.U. Baranger, K. Richter, F. von Oppen, and R.A. Jalabert, unpublished (1999).

314. *Electron–electron Interactions in Disordered Systems,* ed. by A.L. Efros and M. Pollak (North-Holland, Amsterdam, 1985).

315. L.G. Aslamazov and A.I. Larkin, Sov. Phys. JETP **40**, 321 (1975).

316. B.L Altshuler, A.G. Aronov, and A.Yu. Zyuzin, Sov. Phys. JETP **57**, 889 (1983).

317. B.L. Altshuler and A.G. Aronov, in *Electron–electron Interactions in Disordered Systems,* ed. by A.L. Efros and M. Pollak (North-Holland, Amsterdam, 1985).

318. V. Ambegaokar and U. Eckern, Phys. Rev. Lett. **65**, 381 (1990).

319. U. Eckern, Z. Phys. B **42**, 389 (1991).

320. G. Montambaux, J. de Physique **6**, 1 (1996).

321. A. Altland, S. Iida, A. Müller-Groeling, and H.A. Weidenmüller, Ann. Phys. **219**, 148 (1992); Europhys. Lett. **20**, 155 (1992).

322. D. Bouzerar, D. Poiblanc, and G. Montambaux, Phys. Rev. B **49**, 8258 (1994); T. Giamarchi and B.S. Shastry, Phys. Rev. B **51**, 10915 (1995).

323. M. Abraham and R. Berkovits, Phys. Rev. Lett. **70**, 1509 (1993); H. Kato and D. Yoshioka, Phys. Rev. B **50**, 4943 (1994).

324. A. Cohen, K. Richter, and R. Berkovits, Phys. Rev. B **57**, 6223 (1998).

325. M. Ramin, B. Reulet, and H. Bouchiat, Phys. Rev. B **51**, 5582 (1995); H. Kato and D. Yoshioka, Physica B **212**, 251 (1995).
326. F. von Oppen, D. Ullmo, H.U. Baranger, and K. Richter, unpublished (1999).
327. D. Ullmo, unpublished.
328. In the limit where the temperature smoothing is on the order of the mean level spacing, corresponding to a cutoff at the Heisenberg time, off-diagonal terms may play a role in chaotic systems, similar to the effect of such terms in density correlators of noninteracting systems (see Sect. 2.2).
329. T. Dittrich, in [17].
330. M. Brack, O. Genzken, and K. Hansen, Z. Phys. D **21**, 655 (1991).

Index

Printing: Mercedes-Druck, Berlin
Binding: Stürtz AG, Würzburg

Springer Tracts in Modern Physics